Infrared Optical Fibers

The Adam Hilger Series on Optics and Optoelectronics

Series Editors: **E R Pike** FRS and **W T Welford** FRS

Other books in the series
Aberrations in Optical Systems
W T Welford

Laser Damage in Optical Materials
R M Wood

Waves in Focal Regions
J J Stamnes

Laser Analytical Spectrochemistry
edited by V S Letokhov

Laser Picosecond Spectroscopy and Photochemistry of Biomolecules
edited by V S Letokhov

Prism and Lens Making
F Twyman

Cutting and Polishing Optical and Electronic Materials 2nd edition
G W Fynn and W J A Powell

The Optical Constants of Bulk Materials and Films
L Ward

Other titles of related interest

Infrared Optical Materials and their Antireflection Coatings
J A Savage

Principles of Optical Disc Systems
G Bouwhuis, J Braat, A Huijser, J Pasman, G van Rosmalen and
K Schouhamer Immink

Thin-film Optical Filters
H A Macleod

Aberration and Optical Design Theory
G G Slyusarev

Lasers in Applied and Fundamental Research
S Stenholm

The Adam Hilger Series on Optics and Optoelectronics

Infrared Optical Fibers

Toshio Katsuyama

and

Hiroyoshi Matsumura

Central Research Laboratory,
Hitachi Ltd, Tokyo,
Japan

Adam Hilger, Bristol and Philadelphia

© IOP Publishing Ltd 1989

British Library Cataloguing in Publication Data

Katsuyama, Toshio
  Infrared optical fibers.
  1. Infrared fibre optics
  I. Title II. Matsumura, Hiroyoshi
  621.38′0414

  ISBN 0-85274-288-6

Library of Congress Cataloging-in-Publication Data are available

Published under the Adam Hilger imprint by IOP Publishing Ltd
Techno House, Redcliffe Way, Bristol BS1 6NX, England
242 Cherry Street, Philadelphia, PA 19106, USA

Typeset by KEYTEC, Bridport, Dorset
Printed in Great Britain by J W Arrowsmith Ltd, Bristol

To our wives

Mitsuko Matsumura and Tomoko Katsuyama

Contents

Series Editors' Preface

Optics has been a major field of pure and applied physics since the mid 1960s. Lasers have transformed the work of, for example, spectroscopists, metrologists, communication engineers and instrument designers in addition to leading to many detailed developments in the quantum theory of light. Computers have revolutionised the subject of optical design and at the same time new requirements such as laser scanners, very large telescopes and diffractive optical systems have stimulated developments in aberration theory. The increasing use of what were previously not very familiar regions of the spectrum, e.g. the thermal infrared band, has led to the development of new optical materials as well as new optical designs. New detectors have led to better methods of extracting the information from the available signals. These are only some of the reasons for having an *Adam Hilger Series on Optics and Optoelectronics*.

The name Adam Hilger, in fact, is that of one of the most famous precision optical instrument companies in the UK; the company existed as a separate entity until the mid 1940s. As an optical instrument firm Adam Hilger had always published books on optics, perhaps the most notable being Frank Twyman's *Prism and Lens Making*.

Since the purchase of the book publishing company by The Institute of Physics in 1976 their list has been expanded into all areas of physics and related subjects. Books on optics and quantum optics have continued to comprise a significant part of Adam Hilger's output, however, and the present series has some twenty titles in print or to be published shortly. These constitute an essential library for all who work in the optical field.

Preface

Since Pinnow, Van Uitert, Goodman *et al* discussed in 1978 the possibility of an ultra-low loss optical fiber with a loss of less than $0.01 \, \mathrm{dB \, km^{-1}}$, non-silica-based infrared fibers have become a center of interest in optical fiber research. The number of articles appearing in the technical journals has increased significantly, although to our knowledge there have as yet been no books describing infrared optical fibers systematically.

This may be the first book on infrared optical fibers. It is intended as a state-of-the-art review for use by researchers and engineers engaged in these research fields, and also as an introductory textbook for readers wanting to begin research on infrared optical fibers. It is assumed therefore that readers have a basic, but not necessarily extensive, background knowledge of physics and chemistry. Chapters 2 and 3 give the basic concepts of optical fibers for infrared transmission, including the theory of light guiding and of transmission loss. Readers do not therefore need any particular knowledge of optical fibers.

The book is largely influenced by many excellent review papers on the infrared material and fiber technology; particularly I W Donald and P W McMillan, *Review of Infra-red Transmitting Materials* (1978 *J. Mater. Sci.* **13** 1151–76), T Miyashita and T Manabe, *Infrared Optical Fibers* (1982 *IEEE J. Quantum Electron.* **QE-18** 1432–50), and D C Tran, G H Sigel Jr and B Bendow, *Heavy Metal Fluoride Glasses and Fibers: A Review* (1984 *J. Lightwave Technol.* **LT-2** 566–586). In addition, the description of the concepts of optical fibers (§§2.1 and 2.2) is based on the book written by Y Suematsu and K Iga: *Introduction to Optical Fiber Communications* (1982 John Wiley and Sons, New York, translated into English by H Matsumura and W A Gambling). We would like to thank the authors for giving us some ideas on book preparation.

We also thank the editors and authors of some technical journals for allowing us to use figures which appear in this book. The cited journals include *Applied Optics*, *Optics Letters*, the *Journal of Applied Physics*,

Applied Physics Letters, the *IEEE Journal of Quantum Electronics*, the *Journal of Lightwave Technology*, *IEEE Transactions: Microwave Theory and Techniques*, the *Journal of Non-Crystalline Solids*, *Electronics Letters*, the *Materials Research Bulletin*, the *Japanese Journal of Applied Physics*, the *Physics and Chemistry of Glasses*, *Infrared Physics*, the *Journal of Materials Science*, the *Bell System Technical Journal*, *Kougakugijitsu contact* (the *Journal of the Association of Optical and Electro-optical Technology*), *Tsuuken kenkyuu jitsuyouka houkoku (Electrical Communication Laboratories Technical Journal)*, *Denki gakkai zasshi (Journal of Institute of Electrical Engineers of Japan)*, *Ouyou butsuri (Journal of the Japan Society of Applied Physics)* and *Laser kennkyuu (Journal of Laser engineering)*. Figures are also cited from the materials issued by the Institute of Electrical Communication, Tohoku University, Le Verre Fluore and John Wiley and Sons, Inc.

The authors wish to extend their appreciation for the encouragement given by Drs Y Takeda and M Kudo of the Central Research Laboratory of Hitachi Ltd. Thanks are also due to Drs B Yoda, H Nagano and K Mikoshiba of Hitachi Cable Co. for their encouragement throughout infrared fiber research.

Toshio Katsuyama
Hiroyoshi Matsumura
Tokyo, January 1988

1 Introduction

Recently, optical fibers have become a center of interest as transmission lines for such diverse applications as communication links and sensing systems. In particular, optical fiber communications offer an exciting alternative to traditional wire communications. These are mainly based on the successful fabrication of low loss silica-based optical fibers whose transmission losses are reduced to as low as $0.2 \, \text{dB} \, \text{km}^{-1}$. Thus, the low loss quality of silica-based optical fibers has enabled us to construct high bit-rate and long haul communication systems. For example, in Japan optical communication links spanning the Japanese islands have already been established and the telecommunication services using these systems have become commercially available. However, the demand for further improvements in transmission capacity is still increasing. These high capacity optical communication systems essentially require the realization of ultra-low loss optical fibers with losses far below those of the silica-based optical fibers.

Historically Pinnow *et al* (1978), Van Uitert and Wemple (1978) and Goodman (1978) first discussed the possibility of an ultra-low loss, less than $10^{-2} \, \text{dB} \, \text{km}^{-1}$, for infrared materials, and these discussions motivated the research efforts on the non-silica-based infrared optical fibers. Furthermore, there have been increasing demands on laser power transmission through flexible optical fibers in the fields of laser surgery and machining. Since CO_2 laser power transmission is particularly useful in these fields, studies on low loss infrared fibers at a $10.6 \, \mu\text{m}$ wavelength have been extensively performed.

Optical materials studied to date for infrared optical fibers are heavy-metal oxides, halides and chalcogenides. In the heavy-metal oxides, GeO_2-based glasses have been extensively studied since Olshansky and Scherer (1979) predicted a low loss reaching below $0.2 \, \text{dB} \, \text{km}^{-1}$. On the other hand, polycrystalline and single-crystalline halide materials such as TlBr–TlI mixed crystal (which is called KRS-5), AgCl, AgBr, KCl and CsBr have mainly been studied. These crystalline materials are

1

particularly advantageous for laser power transmission because the losses are sufficiently low at a CO_2 laser wavelength of 10.6 μm. Chalcogenides so far studied are basically divided into sulfides, selenides and tellurides whose states are vitreous or glassy. Among them, sulfide glass fibers can transmit light of wavelength between 2 and 5 μm. On the other hand, selenide and telluride glass fibers have a wide transparency range which covers around 10 μm in wavelength. These selenide and telluride glass fibers are therefore being studied for CO_2 (10.6 μm) and CO (5.3 μm) laser power transmissions.

It should be noted that the infrared fiber research has been accelerated by the discovery of ZrF_4-based fluoride glasses by Poulain *et al* (1975). This discovery and the subsequent researches on the ZrF_4-based glasses have made it possible to fabricate low loss infrared fibers. The progress of the loss reduction is so fast that a loss of less than 1 dB km^{-1} has been obtained (Tran 1986, Kanamori and Sakaguchi 1986). These fluoride glasses are therefore thought to be the most promising candidates for the ultra-low loss optical fibers in long distance optical communications. The loss value predicted is less than 0.01 dB km^{-1} at 2–4 μm wavelengths.

On the other hand, various hollow waveguides have been studied mainly for infrared light power transmission, particularly CO_2 laser power transmission at a 10.6 μm wavelength. The transmission characteristics of various metallic hollow waveguides, such as parallel-plate metallic waveguides and dielectric-coated metallic hollow waveguides, have been improved. Hollow core fibers utilizing total reflections between the hollow cores and dielectric claddings have also been proposed. This light guiding is possible only in the wavelength regions of abnormal dispersion where the refractive indices become lower than unity.

Applications of infrared optical fibers can be classified into two categories: long distance optical communications and short haul light transmissions, as shown in the above description. Among them, long distance optical communications require infrared fibers which have ultra-low losses and low dispersions. On the other hand, short haul light transmissions require them with wide band transparency and/or high power light transmission. The applications in short haul transmissions are, for example, optical transmissions in a nuclear radiation environment, infrared remote sensing such as temperature measurement by thermal radiations, and laser surgery and machining.

This book is divided into eight chapters. Following this introductory chapter, basic concepts of optical fibers including the mechanism of light guiding, refractive index and dispersion properties, and transmission loss characteristics are described in chapter 2. These concepts are fundamental to the understanding of optical fibers. In chapter 3, introductory remarks on infrared optical fibers are presented. The historical sketch, materials for infrared transmission, classifications and applications of the infrared optical fibers studied to date, and some measurement techniques of transmission properties are briefly described. Thus one can obtain an overview of the infrared optical fiber researches and their applications.

Chapters 4 and 5 are devoted to the detailed descriptions of the glass fibers and crystalline fibers for infrared transmission. Materials, fabrication techniques and properties of these fibers are described. A hollow waveguide whose light guiding mechanism is slightly different from the conventional optical fibers is presented in chapter 6. Furthermore, chapter 7 describes in detail the applications of infrared optical fibers, including ultra-long repeaterless links, nuclear radiation resistant links, measurement systems of thermal radiation, and laser power transmission systems. Finally, concluding remarks and prospects for future work are added in chapter 8.

References

Goodman C H L 1978 *Solid-State Electron Dev.* **2** 129–37

Kanamori T and Sakaguchi S 1986 *Japan. J. Appl. Phys.* **25** L468–70

Olshansky R and Scherer G W 1979 *Proc. 5th European Conf. on Optical Communication and 2nd Int. Conf. on Integrated Optics and Optical Fiber Communication* 12.1.1–12.5.3

Pinnow D A, Gentile A L, Standlee A G, Timper A J and Hobrock L M 1978 *Appl. Phys. Lett.* **33** 28–9

Poulain M, Poulain M, Lucas J and Brun P 1975 *Mater. Res. Bull.* **10** 243–6

Tran D C 1986 presented at *Conf. on Optical Fiber Communication (Atlanta)*

Van Uitert L G and Wemple S H 1978 *Appl. Phys. Lett.* **33** 57–9

2 Basic Concepts of Optical Fibers

In this chapter the fundamental properties of optical fibers are described. The mechanism of light guiding, refractive index and dispersion properties, and transmission loss characteristics are the most important concepts for understanding optical fibers, including infrared optical fibers.

2.1 The mechanism of light guiding

2.1.1 The basic structure of an optical fiber
An optical fiber consists of a central part called the 'core', surrounded by a material called the 'cladding', as shown in figure 2.1. The core has a refractive index n_1, which is higher than that of the cladding n_2. Therefore, electromagnetic waves can be confined in the core region and are transmitted by total internal reflections at the boundary between the core and cladding. In the low loss silica glass fiber which is used for optical fiber communications the typical core diameter is usually in the range from a few micrometers to tens of micrometers, and the outer cladding diameter is fixed at 125 μm. Also optical fibers are normally given a plastic primary coating and then a nylon coating, because fibers without coating are very weak mechanically and are subject to chemical attacks by moisture. An optical fiber cable may contain several coated optical fibers.

2.1.2 Refraction and reflection of a light ray
Since the light is described as an electromagnetic wave, its propagation property must be explained in terms of Maxwell's equations. However,

it is often convenient to use geometrical optics when the wavelength of light is considerably shorter than the core dimension.

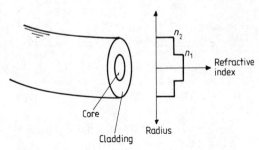

Figure 2.1 The structure of an optical fiber.

First let us consider the refraction and reflection of light at the boundary of two dielectric media with different refractive indices. These are most important basic concepts in light guiding.

In figure 2.2, n_1 and n_2 are the refractive indices of media I and II respectively. In the case of $n_1 < n_2$ (figure 2.2(a)), the light ray, which is projected obliquely onto the boundary from the upper left, changes direction at the boundary. The refraction angle follows from Snell's law, that is

$$\frac{\sin \alpha_1}{\sin \alpha_2} = \frac{n_2}{n_1}, \tag{2.1}$$

where α_1 and α_2 are the angles made by the input and refracted rays, respectively, with the normal to the interface. For the complementary angles θ_1 and θ_2, equation (2.1) can be expressed as

$$\frac{\cos \theta_1}{\cos \theta_2} = \frac{n_2}{n_1}. \tag{2.2}$$

On the other hand, when $n_1 > n_2$ (figure 2.2(b)), then θ_2 decreases with decreasing θ_1, until finally we find $\theta_2 = 0$ at a certain finite value of θ_1, because $\theta_1 > \theta_2$ from equation (2.2). At this point, transmission of the light wave into medium II ceases completely, and all the energy is totally reflected at the boundary. The angle $\theta_1 = \theta_c$ at which this total reflection occurs is given, from equation (2.2), by

$$\theta_c = \cos^{-1} \left(\frac{n_2}{n_1} \right). \tag{2.3}$$

θ_c is called the 'critical (complementary) angle' or the 'total reflection complementary angle'.

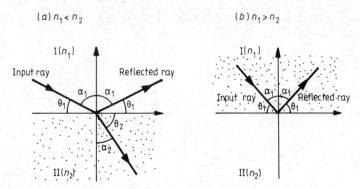

Figure 2.2 Refraction and reflection of a light ray, for (*a*) $n_1 < n_2$, and (*b*) $n_1 > n_2$ (total reflection).

2.1.3 *The mechanism of light guiding*

Let us consider the principle of the guiding mechanism of the optical fiber. The refractive index n_1 in the central region (called the core) in figure 2.3 is higher than n_2, that of the surrounding region (called the cladding). Furthermore, the refractive index in the core is uniform, forming the 'step-index optical fiber'. Consider a ray 1 in air at an angle θ' to the fiber axis, as in figure 2.3, striking the core. Due to refraction at the air/fiber surface, the angle of the ray to the axis changes to θ as it enters the core, where, from equation (2.1),

$$\frac{\sin \theta'}{\sin \theta} = n_1, \tag{2.4}$$

since the refractive index of air is 1.

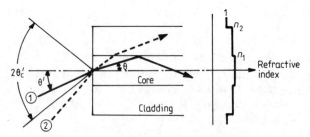

Figure 2.3 Light transmission in an optical fiber.

If the angle θ of the ray to the axis inside the core is smaller than the total reflection complementary angle $\theta_c = 90° - \alpha_c = \cos^{-1} (n_2/n_1)$

(where α_c is the critical angle), then complete reflection occurs and the ray continues to propagate along the core, since all subsequent reflections occur at the same angle and therefore with no loss of energy. On the other hand, if a ray 2 enters at such a wide angle that inside the core it strikes the core/cladding boundary at an angle greater than θ_c, then only partial reflection takes place and some of the energy is lost by refraction into the cladding. After several successive reflections, very little energy is left in the core and the guidance is lost. Thus only those rays up to an angle θ'_c in air are accepted and guided by the core.

In most practical situations, and especially with optical fibers, the difference between n_1 and n_2 is small, that is $n_1 - n_2 \ll n_1$, so that the 'relative refractive index difference' Δ can be defined by

$$\Delta = \frac{n_1^2 - n_2^2}{2n_1^2} \simeq \frac{n_1 - n_2}{n_1}. \tag{2.5}$$

Δ is usually expressed as a percentage.

The total reflection complementary angle in the core from equation (2.3) can be written in terms of Δ as follows:

$$\theta_c = \cos^{-1}(n_2/n_1) = \cos^{-1}(1 - \Delta) = 2\sin^{-1}(\tfrac{1}{2}\Delta)^{1/2} \tag{2.6}$$

$$\simeq (2\Delta)^{1/2} \qquad \Delta \ll 1.$$

Thus, when $\Delta = 1\%$ we have $\theta_c = 0.14$ rad $= 8.0°$.

The maximum acceptance angle (figure 2.3) is given by $2\theta'_c$, where $\theta'_c = \sin^{-1}(n_1 \sin \theta_c)$, so that

$$2\theta'_c = 2\sin^{-1}(n_1 \sin \theta_c) \simeq 2\sin^{-1}(n_1^2 - n_2^2)^{1/2}. \tag{2.7}$$

Another important definition is that of 'numerical aperture', often abbreviated to NA, which is given by

$$NA = \sin \theta'_c = n_1 \sin \theta_c = (n_1^2 - n_2^2)^{1/2} \simeq n_1 (2\Delta)^{1/2}. \tag{2.8}$$

Thus if $\Delta = 1\%$ and $n_1 = 1.5$, then $NA = 0.21$ and $2\theta'_c = 24°$.

The parameters such as the relative refractive index difference Δ, the acceptance angle $2\theta'_c$ and the numerical aperture NA are the fundamental values describing the characteristics of optical fibers.

2.1.4 The concept of modes

An optical wave guided by successive total internal reflections may be represented by bundles of rays called the 'modes'. By using the idea of the mode, the phenomenon of light guidance, so far expressed by the total internal reflections of rays, can also be treated as a wave, which is the other characteristic of light. In order to provide a basic explanation of the modes, we consider for simplicity a two-dimensional slab of uniform material. Figure 2.4 shows the rays which make an angle $\pm\theta$

with the core/cladding interface of the slab waveguide. Each ray represents a plane wave from the wave point of view and is drawn perpendicularly to the wavefront.

The quantity

$$k_0 = \frac{2\pi}{\lambda} \tag{2.9}$$

denotes the 'phase constant' and λ the wavelength of the plane wave in a vacuum. Within the core, where the refractive index is n_1, the wavelength becomes smaller $(= \lambda/n_1)$ while the phase constant is larger $(= k_0 n_1)$.

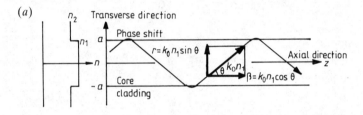

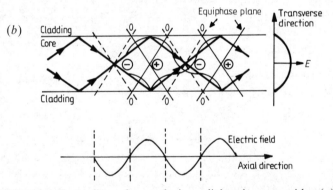

Figure 2.4 Formation of a mode in a dielectric waveguide. (*a*) Decomposition of the propagation direction. (*b*) Interference of the incident and reflected waves (after Suematsu and Iga 1982).

This plane wave can be resolved into two component plane waves propagating in the axial and transverse directions, as shown in figure 2.4(*a*). The plane wave, or equivalent ray, travels along its path with a phase constant $k_0 n_1$ and at a constant angle $\pm\theta$ to the axial direction by successive reflections along the complete length of the waveguide. The component of the phase constant in the axial direction is $k_0 n_1 \cos \theta$, and this is also, therefore, the axial propagation constant β. That is

$$\beta = k_0 n_1 \cos \theta. \tag{2.10}$$

On the other hand, if γ denotes the transverse component of the propagation constant, it is similarly given by

$$\gamma = \pm k_0 n_1 \sin \theta. \tag{2.11}$$

The transverse component of the plane wave is reflected at the core/cladding interfaces, so that when the total phase change after two successive reflections at the upper and lower interfaces becomes $2m\pi$ (where m is any integer), a standing wave is established in the transverse direction. Figure 2.4(b) shows this self-consistent field in terms of the interference of the two light rays, where a positive electric field direction along the phase plane vertical to the ray is expressed by the faint solid lines and the negative field by the faint dashed lines.

Near the boundary between the core and the cladding, the positive and negative phase planes always coincide, so that the electric field becomes zero. On the other hand, in the central region the fields sum and the combined electric field on the axis becomes large. This behaviour is equivalent to the confinement of the optical wave.

In the case just described, the field distribution in the transverse direction does not change as the wave propagates in the axial direction. This kind of stable field distribution is called the mode and is obtained only when the angle between the ray and the interface has a particular value.

Figure 2.5 shows the field distribution for each mode. The mode is characterized by the mode number, which is equal to the number of zeros of the intensity distribution in the transverse direction. This property of the mode is, of course, applicable to the slab waveguide. However, the propagation characteristics of a circular cylindrical fiber can be easily predicted by using a certain correspondence between the slab and the cylindrical fiber.

So far we have considered the so-called multimode optical fiber. The multimode optical fiber can support a large number of modes. In describing the characteristics of the circular fiber, it is convenient to introduce the normalized frequency V:

$$V = \frac{2\pi}{\lambda} n_1 a (2\Delta)^{1/2} = (n_1^2 - n_2^2)^{1/2} k_0 a, \tag{2.12}$$

where λ is the free-space wavelength and a is the core radius. The multimode operation results when $V > 2.4$. In contrast, the optical fiber with a normalized frequency of less than 2.4 is called the 'single-mode optical fiber'. The single-mode optical fiber supports only one fundamental mode.

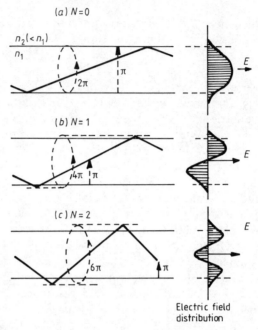

Figure 2.5 Ray propagation and electric field distribution of modes in a step-index waveguide. N represents the mode number (after Suematsu and Iga 1982).

2.2 Dispersion characteristics

2.2.1 Transmission bandwidth of optical fiber

The transmission bandwidth of an optical fiber restricts the amount of information that can be transmitted through the fiber. An important factor determining the bandwidth is the variation of group velocity with wavelength and the fact that it changes from one mode to another. The group velocity v_g is given in general by

$$v_g = \left(\frac{\partial \beta}{\partial \omega}\right)^{-1}, \tag{2.13}$$

where β is the propagation constant and ω $(= 2\pi f)$ is the angular frequency.

The propagation constant changes with frequency f because of

(i) the frequency dependence (and wavelength dependence, since $c = f\lambda$) of the refractive index n_1 in the core (through the material dispersion parameter $(\lambda/c)(d^2 n/d\lambda^2)$), and

(ii) the frequency dependence of the propagation constant, which is a function of the waveguide structure, that is, the waveguide dispersion.

These factors are present in each individual mode, but in multimode fibers they are generally swamped by the spread in group velocities of the various propagation modes, normally referred to as intermode dispersion.

The propagating time of a signal over a distance L is given by

$$\tau = \frac{L}{v_g} = L\frac{\partial \beta}{\partial \omega} \qquad (2.14)$$

and is called the group delay time. Assuming that the broadenings of the group delay times due to 'material dispersion', 'waveguide dispersion' and 'intermode dispersion' are $\Delta\tau^{(n)}$, $\Delta\tau^{(g)}$, and $\Delta\tau^{(m)}$, respectively, then the total group delay is given approximately by

$$\Delta\tau = \Delta\tau^{(n)} + \Delta\tau^{(g)} + \Delta\tau^{(m)}. \qquad (2.15)$$

In multimode fibers the intermode dispersion generally predominates, and the relative values are such that

$$\Delta\tau^{(m)} \gg \Delta\tau^{(n)} > \Delta\tau^{(g)}. \qquad (2.16)$$

$\Delta\tau^{(m)}$ depends on the refractive index distribution in the core and on the amount of mode coupling and mode filtering that take place during propagation. This term is absent in single-mode fibers, which therefore have a much larger bandwidth.

2.2.2 Material dispersion

In multimode fibers the propagating wave is largely confined to the core so that the material dispersion is nearly the same as that of the bulk core glass. Thus the propagation constant is given, for this particular case, as $\omega n_1/c$, and since n_1, as well as ω, is a function of λ, the transmitted pulse is broadened. Assuming that the group delay broadening due to this refractive index dispersion is $\Delta\tau^{(n)}$ and the wavelength spread of the light source is $\Delta\lambda$, then $\Delta\tau^{(n)}$ can be obtained from equation (2.14) as

$$\Delta\tau^{(n)} \simeq \frac{\partial}{\partial \lambda}\left(\frac{\partial \beta}{\partial \omega}\right) L\Delta\lambda = -L\lambda\left(\frac{\lambda}{c}\right)\left(\frac{d^2 n_1}{d\lambda^2}\right)\left(\frac{\Delta\lambda}{\lambda}\right). \qquad (2.17)$$

Here $-(\lambda/c)(d^2 n_1/d\lambda^2)$ represents the delay time per unit wavelength spread and unit transmission distance and is referred to as the material dispersion coefficient. For positive values of $d^2 n_1 / d\lambda^2$, the longer wavelength components of $\Delta\lambda$ travel faster than the shorter ones, and vice versa.

Fused silica glass, which is widely used as the core material of conventional optical fibers, has the smallest material dispersion in the interesting wavelength region. In figure 2.6, its material dispersion is shown as a function of wavelength. It can be seen from the figure that the material dispersion is just zero around $1.3\,\mu$m. Therefore, this critical wavelength is called the zero material dispersion wavelength λ_c. In the wavelength of λ_c, a larger transmission bandwidth can be obtained when the single-mode fibers may be used. Each value of the material dispersion for infrared optical fibers will be described in detail in chapters 4–6.

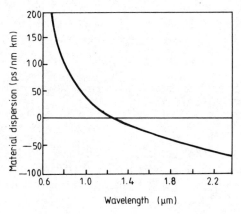

Figure 2.6 The material dispersion $(\lambda/c)(d^2 n / d\lambda^2)$ of silica glass. The group delay spread for 1 nm spectral width and 1 km fiber length is given as a function of wavelength (after Payne and Gambling 1975).

2.2.3 Waveguide dispersion

The waveguide dispersion $\Delta\tau^{(g)}$ in a fiber results from the nonlinearity of the propagation constant with the optical frequency.

The waveguide dispersion in the $0.85\,\mu$m wavelength region of conventional silica glass fibers is such that the group velocity is larger at longer wavelengths. This is the same as for material dispersion, so the two dispersions are additive, but the waveguide dispersion is about one order of magnitude smaller than material dispersion. However, the material dispersion is extremely small at a $1.25\,\mu$m wavelength, so the two dispersions become comparable at this wavelength. Furthermore at about 1.27 μm they have different signs and cancel each other out, so that the combined effect becomes zero.

2.2.4 Intermode dispersion

In the case of the multimode fiber, in which a large number of modes can propagate, the group velocities of the modes are different, so that a pulse propagating in such a fiber is broadened. This is shown schematically in figure 2.7. At the input end a large number of modes is, in general, excited, each with its own propagation velocity. The detector at the output end responds to the total power at any instant so that the output signal is spread out in time. Even if only a single mode is excited at the input, then, after some distance, mode conversion (and reconversion) effects due to inhomogeneities in the core can cause energy to be scattered into other modes.

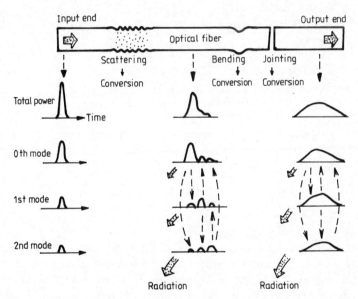

Figure 2.7 Mode dispersion and mode coupling in an optical fiber (after Suematsu and Iga 1982).

Among the various optical fibers, multimode step-index optical fibers show the largest intermode dispersion. On the other hand, single-mode optical fibers show the smallest intermode dispersion because only one mode can propagate.

2.3 Transmission loss characteristics

2.3.1 General description of transmission loss

The intensity of transmitted light gradually decreases with the distance because of absorptions and scatterings. The intensity I_t of the light transmitted in the material of thickness t is expressed as

$$I_t = I_i \exp(-\alpha t), \tag{2.18}$$

where I_i is the intensity of the input light and α is the 'absorption coefficient'. Equation (2.18) is called the Lambert–Bouger law.

If the reflections at the input and output surfaces are taken into account, equation (2.18) can be rewritten as

$$I_t = I_i(1 - R)^2 \exp(-\alpha t), \tag{2.19}$$

where R is the reflectivity with air at normal incidence, and is approximately given by

$$R \simeq \left(\frac{n-1}{n+1}\right)^2. \tag{2.20}$$

Here, n is the refractive index of the material. The insertion of equation (2.20) into equation (2.19) gives

$$\frac{I_t}{I_i} \simeq \left\{1 - 2\left(\frac{n-1}{n+1}\right)\right\}^2 \exp(-\alpha t). \tag{2.21}$$

The absorption coefficient α is normally expressed in cm^{-1}. The relation between the absorption coefficient α (in cm^{-1}) and the 'transmission loss' L (in $dB\,m^{-1}$) is shown as

$$L[dB\,m^{-1}] = 434\alpha[cm^{-1}]. \tag{2.22}$$

2.3.2 The transmission loss mechanism

The transmission loss of the optical fiber consists of intrinsic and extrinsic losses. The intrinsic losses are the intrinsic absorption loss and the scattering loss, while the extrinsic losses are the impurity absorption loss and the scattering loss due to structural imperfection. These are listed in table 2.1 and are shown schematically in figure 2.8.

2.3.2.1 Intrinsic absorption loss

Loss due to electronic transition. In silica glass, the absorption edge caused by electronic transition is located around 0.16 μm. In general, this absorption is very small in the visible and infrared regions. The tail of this absorption in silica glass is estimated to be less than 0.2–

Table 2.1 Loss factors of optical fibers.

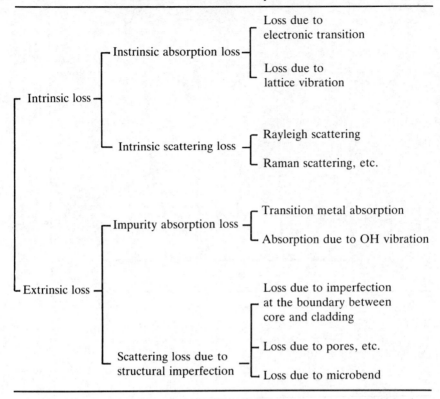

0.3 dB km^{-1} at a wavelength of 0.8 μm and becomes smaller for wavelengths longer than 1 μm.

The absorption coefficient α due to electronic transition is approximately expressed as

$$\alpha = \exp{(E - E_g)}/E_0, \qquad (2.23)$$

where E is the photon energy $h\nu$ (h is Planck's constant and ν the frequency of the light), E_g is the energy gap and E_0 is constant.

Loss due to lattice vibration. The loss due to lattice vibration depends on the vibrational modes of the constituent atoms and molecules. When the frequency of the incident light is the same as that of the intrinsic lattice vibration, the light is strongly absorbed by the material. The molecule, then, must possess a permanent dipole moment which can be activated by the oscillating electric field of the incident light. This first

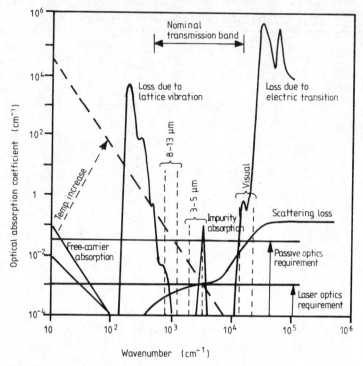

Figure 2.8 Various losses of optical fibers (after Miles 1976).

order effect occurs in ionic crystals only, and is known as 'reststrahlen' absorption.

Vibrations of linear non-polar molecules (i.e. O_2, N_2, etc.) cause no change in electric moment and so these molecules would not be expected to exhibit infrared vibrational spectra. However, absorption may be observed under certain conditions because of induced dipole moments. One vibrational mode (which is infrared inactive) may induce charges on the atoms, and a second mode may simultaneously cause a vibration of the induced charges, so producing an electric moment of the second order, and hence coupling to the field of the incident light.

The absorption coefficient for these second order effects is generally several orders of magnitude smaller than reststrahlen absorption, but may still effectively render a material opaque over a given wavelength region.

The simplest model of vibration of a linear diatomic molecule consisting of two point masses, m_1 and m_2, is a simple harmonic oscillation of the two masses. This is the classical description of a linear polar diatomic molecule, and the frequency of the vibration $\bar{v}$, according to this model, is given by

$$\bar{v} = \frac{1}{2\pi}(f/\mu)^{1/2} \qquad \frac{1}{\mu} = \frac{1}{m_1} + \frac{1}{m_2}, \qquad (2.24)$$

where μ is the reduced mass and f the restoring force, or force constant.

Such a simple equation is very effective in predicting the fundamental absorption frequency of many polar diatomic molecules, as well as some multiatomic systems, and even some homopolar systems. It can be predicted that when the frequency $\bar{v}$ is small, the absorption band exists in the longer wavelength region. Therefore, it can be easily understood that $\bar{v}$ must be small in order to obtain an excellent transparency in the infrared region. This situation corresponds to the small force constants and the large masses of the constituent atoms.

It should be noted that the frequency of equation (2.24) expresses the fundamental mode. However, there exists a first overtone and, of course, higher overtones, but the absorption intensities of these overtones are small compared to that of the fundamental vibration.

2.3.2.2 Intrinsic scattering loss

The most important intrinsic scattering loss is the Rayleigh scattering loss. Other scattering losses (e.g. from Raman scattering, Mie scattering, etc.) are relatively small.

Rayleigh scattering is caused by the small particles or inclusions in the material. It can be shown that the angular intensity distribution of radiation scattering by second-phase inclusions which are small compared to the wavelength of the incident light is of the form

$$I(\theta) = \left(\frac{1 + \cos^2\theta}{x^2}\right) \frac{8\pi^4}{\lambda^4} r^6 \left|\frac{N^2 - 1}{N^2 + 1}\right|^2 I_0, \qquad (2.25)$$

where $I(\theta)$ is the specific intensity at the scattering angle θ, x is the distance from the scattering center, r is the radius of the inclusion, N is the ratio of refractive indices of inclusion and medium, and I_0 is the incident intensity.

When the refractive index of the inclusion is larger than that of the medium, the scattering light intensity changes linearly with r^6/λ^4 and $|(N^2 - 1)/(N^2 + 1)|^2$. Therefore, in order to reduce the scattering loss, it is required that the refractive index of the inclusion must be roughly equal to that of the medium and the radius of the inclusion must be small.

Equation (2.25) is a general formula which can be used for both crystals and glasses. Here, discussion is concerned only with glass.

In general, when glass solidifies, a density fluctuation, that is a refractive index fluctuation, results from the inhomogeneity caused by slight discrepancies in cooling rate. This is one origin of the scattering loss.

The scattering loss coefficient α_s of the glass composed of only one element is given by

$$\alpha_s = \frac{8}{3} \frac{\pi^3}{\lambda^4} (n^8 p^2)(kT) \beta_T, \qquad (2.26)$$

where n is the refractive index of the medium, p is the photo-elastic coefficient, k is the Boltzmann constant, T is the glass transition temperature (the absolute temperature) and β_T is the isothermal compressibility (Pinnow *et al* 1973).

Equation (2.26) shows that the scattering loss coefficient α_s is proportional to λ^{-4}. This means that the amount of scattered light increases as the wavelength decreases. It also shows that optical glass materials with a lower refractive index and a lower softening temperature will give smaller scattering losses.

On the other hand, in the case of binary or ternary glass systems, there is also the scattering loss due to the inhomogeneity of composition. This scattering loss is also Rayleigh scattering loss, and is given by

$$\alpha_{sc} = \frac{16\pi^3 N}{3\lambda^4} \left(\frac{2N}{2C}\right)^2 \overline{\Delta C^2} \, \delta V, \qquad (2.27)$$

where C is the density, $\overline{\Delta C^2}$ is the mean square of the density fluctuation, and δV is the volume of the inhomogeneous region.

2.3.2.3 Impurity absorption

Impurity absorption loss is caused mainly by impurities introduced during the fabrication process. In the silica glass fibers, transition metals such as Fe and Co, and OH cause a strong absorption in the relevant wavelength region. A reduction in the level of these impurities is necessary if light is to be transmitted through long distances. Figure 2.9 shows the absorption losses per unit ppb of various transition metals in the silica glass fibers. It can be seen from this figure that an order of 1 ppt (10^{-12}) impurities are necessary to fabricate low loss optical fiber.

On the other hand, the absorption due to OH impurity corresponds to the resonant absorption of the vibrational mode in the infrared range. The loss per unit ppm of OH bonds is shown in figure 2.10. In the silica glass fibers, it is known that the fundamental absorption of Si–OH vibration exists at a 2.73 μm wavelength. The absorptions due to the combination with Si–O vibrations also appear at the wavelengths of 1.23 and 1.98 μm. It is clear that the reduction of OH impurity levels is very important for transmitting light of wavelength longer than 1.2 μm.

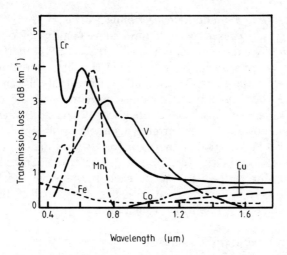

Figure 2.9 Absorption losses per unit ppb of various transition metals in silica glass fibers (after Schultz 1974).

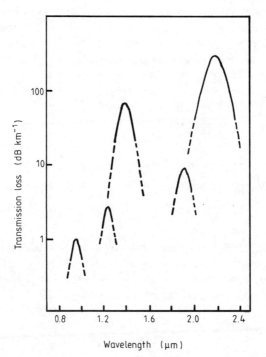

Figure 2.10 Transmission loss per unit ppm of OH bonds in silica glass fibers (after Osanai 1980).

2.3.2.4 Scattering loss due to structural imperfection

The scattering loss caused by structural imperfections of optical fibers does not depend on the wavelength. Typical structural imperfections are the boundary irregularity between core and cladding, pores in the core, and microbend of the fiber. The boundary irregularity is caused by the discrepancy in the physical and chemical properties of the core materials. In particular, the discrepancy in viscosity and softening temperature plays an important role in forming a smooth boundary.

It should also be noted that the variation of core diameter along the fiber length is caused by the variation of the outer diameter of the fiber, which appears during the drawing process. It is known, however, that the loss due to core diameter variation is negligibly small if the variation of the outer diameter is controlled to within ± 1 μm.

2.3.3 Formulation of transmission loss

In this section, the formulation of the various transmission losses mentioned in the previous section is described.

First of all, the absorption due to electronic transition can be represented from equation (2.23) as $C \exp(D/\lambda)$, where λ is the wavelength, and C and D are the constants. Furthermore, we represent the absorption due to lattice vibration and the impurity absorption as $E(\lambda)$ and $F(\lambda)$, respectively. Of the intrinsic scattering losses, the Rayleigh scattering loss is dominant, and the other scattering losses are negligibly small. Therefore, the intrinsic scattering loss can be shown as A/λ^4. The scattering loss due to structural imperfection does not depend on the wavelength and is therefore a constant, B .

Consequently, the transmission loss α of the optical fiber can be given by

$$\alpha = A/\lambda^4 + B + C \exp(D/\lambda) + E(\lambda) + F(\lambda). \qquad (2.28)$$

In silica glass fibers, the transmission loss α can be shown as

$$\alpha = \frac{A}{\lambda^4} + B, \qquad (2.29)$$

because the losses due to electronic transition, lattice vibration and impurity absorption are relatively small for wavelengths between 0.6 and 1.5 μm. Therefore, we can easily obtain the values of parameters A and B by fitting the experimental results to equation (2.29).

References

Miles P A 1976 *Opt. Engng* **15** 451–9
Osanai H 1980 *J. Inst. Electronics and Communication Engineers of Japan* **63** 385–95 (in Japanese)

Payne D N and Gambling W A 1975 *Electron. Lett.* **11** 176–8
Pinnow D A, Rich T C, Ostermayer F W and Didomenico J 1973 *Appl. Phys. Lett.* **22** 527–9
Schultz P C 1974 *J. Am. Ceramic Soc.* **57** 309–13
Suematsu Y and Iga K 1982 *Introduction to Optical Fiber Communications* (New York: Wiley)

3 Introduction to Infrared Optical Fibers

This chapter introduces infrared optical fibers, and reviews their classification and applications. The chapter also provides an overview of infrared fiber research, and discusses techniques for measuring transmission properties.

3.1 A historical sketch

Infrared optical fiber fabrication was first announced by Kapany and Simms in 1965. They fabricated As_2S_3 chalcogenide glass fibers by means of a double crucible method. These fibers were intended to be applied to short length light transmission such as image guiding and infrared remote sensing since the transmission losses remained more than $10000\,dB\,km^{-1}$ (Kapany and Simms 1965). This work, however, did not lead to greater activity in infrared optical fiber research. Instead, rapid advances have been made in silica-based optical fibers since the successful fabrication method of low loss silica-based fibers was proposed by Kapron $et\,al$ (1970). The transmission loss of as low as $0.2\,dB\,km^{-1}$ was obtained after the extensive studies on fabrication techniques such as impurity reduction and improvement of glass homogeneity (Miya $et\,al$ 1979). This value is the ultimate intrinsic loss for a silica-based fiber, so new infrared optical fiber materials were in turn required for the further reduction of the fiber transmission loss. These materials had to possess infrared absorption edges of longer wavelengths and smaller Rayleigh scattering losses than the silica-based glasses.

In 1978, Pinnow $et\,al$ (1978), Van Uitert and Wemple (1978) and Goodman (1978) indicated, for the first time, the possibility of ultra-low loss, less than 10^{-2} dB km^{-1}, for infrared optical transmitting materials, and this motivated research efforts on the non-silica-based infrared fiber materials.

The infrared optical materials so far studied have been heavy-metal oxides, halides and chalcogenides. In the heavy-metal oxides, GeO_2-based glasses have been most extensively studied since Olshansky and Scherer (1979) predicted a low loss reaching below 0.2 dB km^{-1}. A loss of 4 dB km^{-1} at 2 μm was obtained (Takahashi and Sugimoto 1984). However, it seems to be difficult to reduce the loss further because the elimination of the impurities is difficult compared to the silica-based glasses.

In the halide materials, polycrystals and single crystals such as TlBr–TlI mixed crystal (which is called KRS-5), AgCl, AgBr, KCl and CsBr have mainly been studied. These crystals are particularly advantageous for laser power transmission because the losses are sufficiently low at a CO_2 laser wavelength (10.6 μm). Sakuragi *et al* (1981) succeeded for the first time in transmitting a high power CO_2 laser light by using a TlBr–TlI crystalline fiber. Since then a large number of institutes have tried to improve their transmission characteristics. The realization of truly applicable fibers, however, requires the suppression of additional losses which are due to plastic deformation occurring in crystalline materials.

The chalcogenides studied to date are basically divided into sulfides, selenides and tellurides which are in a vitreous or glassy state. Sulfide glass fibers can transmit light of wavelength 2–5 μm. The transmission loss has been reduced to less than 0.1 dB m^{-1} (Kanamori *et al* 1985). However, further loss reduction is thought to be difficult because of the existence of the intrinsic weak absorption tail. On the other hand, selenide and telluride glass fibers have a wide transparency range, which reaches up to around 10 μm in wavelength. These glass fibers are therefore being studied for CO_2 and CO laser power transmissions (Katsuyama and Matsumura 1986).

It should also be noted that infrared optical fiber research has been accelerated by the discovery of ZrF_4-based fluoride glasses by Poulain *et al* (1975). This discovery and subsequent research on the ZrF_4-based glasses have made possible the fabrication of infrared optical fibers. The progress in loss reduction has been so fast that low losses of less than 1 dB km^{-1} have been obtained (Tran 1986, Kanamori and Sakaguchi 1986). These fluoride glasses are therefore thought to be the most promising candidate for the ultra-low loss optical fibers in long distance optical communications. Figure 3.1 shows the rates of loss reduction of various infrared optical fibers. In the figure, losses of conventional silica-based fibers are also shown for comparison. Significant advances in loss reduction can be recognized from the figure, although the minimum losses of the infrared optical fibers still remain higher than that of silica-based fiber.

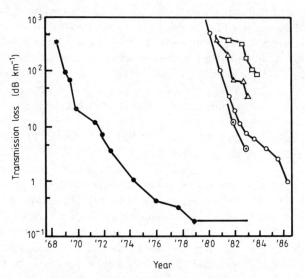

Figure 3.1 The history of loss reduction of various infrared optical fibers (after Yoshida 1986). ●, silica; ⊙, heavy-metal oxides; ○, fluoride glasses; △, chalcogenide glasses; □ halide crystals.

Various hollow waveguides have been studied mainly for infrared light power transmission, particularly CO_2 laser power transmission at a 10.6 μm wavelength. Such waveguides were first proposed by Nishihara *et al* (1974). Garmire *et al* (1976) studied extensively the transmission properties of parallel-plate metallic hollow waveguides. Higher than 200 W c.w. CO_2 laser power has been transmitted through the parallel-plate waveguides (Garmire *et al* 1979). On the other hand, dielectric-coated metallic hollow waveguides developed by Miyagi *et al* (1983), whose transmission properties are improved by introducing dielectric coatings on the inner surfaces of the metallic waveguides, seem to offer a stable light power transmission. However, these waveguides are not as flexible as conventional optical fibers, and so their applications are somewhat limited.

Hidaka *et al* (1982) proposed hollow fibers utilizing total internal reflection between the hollow core and dielectric cladding. This light guiding is possible only in the wavelength regions of abnormal dispersion where the refractive indices are lower than unity. Although a low loss has not yet been attained, the predicted loss is less than 0.1 dB m^{-1} at a 10.6 μm wavelength.

3.2 Materials for infrared transmission

3.2.1 A review

Infrared transmitting materials were mainly used as window materials for infrared spectroscopy in the early stages of research, although recently new materials have been developed for the transmission of high energy laser lights. In this section, the history of infrared material research is briefly reviewed.

Infrared transmitting materials were first studied systematically by Melloni (1833a,b), who reported the kinds of materials which are transparent to the radiation from heated bodies. The materials studied were $NaCl$, SiO_2, CaF_2, and so on. These materials played an important role in the research fields of infrared physics and infrared spectroscopy. In the 1920s, fabrication processes were rapidly improved and large halide crystals became available (Stöber 1925, Strong 1930). As a result, infrared materials were widely used as window materials for infrared transmission.

CaF_2 crystals and thallium halide materials called KRS-5 (Joos 1948) and KRS-6 (Hettner and Leisegang 1948) were artificially synthesized for military use. During 1957–1967, hot-pressed polycrystals composed of MgF_2, ZnS, CaF_2, $ZnSe$, MgO, and $CdTe$ were developed. In particular, MgF_2 polycrystals were found to have an excellent infrared transparency and were named Irtran-1 by the Eastman Kodak Company (Parsons 1972). Since then semiconductor materials such as Ge and Si have also been used as infrared materials.

Chalcogenide glasses (Hilton *et al* 1966), single-crystalline Al_2O_3, polycrystalline Yttralox (Y_2O_3/ThO_2), and PLZT ceramics (which is an electro-optical material (Land *et al* 1974)) have recently been developed. Fabrication techniques have also been improved, resulting in increased strength of the alkali halides by forging (Harrison 1975), and increased volume of $ZnSe$ by chemical vapor deposition (P Miles 1976).

3.2.2 Classification of infrared transmitting materials

Infrared transmitting materials have, in general, small force constants and large masses of constituent atoms, as shown in equation (2.24). The materials are therefore relatively soft and usually contain heavy metals.

Infrared transmitting materials can basically be classified into crystalline and glass forms. In this classification, the former includes polycrystalline forms. Crystalline forms consist of oxides, halides, semiconducting elements, chalcogenides and intermetallic compounds. Glass forms are classified into oxides, halides and chalcogenides. Typical materials for each category are summarized in table 3.1, and table 3.2 shows the regions of the periodic table which provide components for infrared

Table 3.1 Classification of infrared materials.

Form	Group	Material
Crystal	Oxides	SiO_2, Al_2O_3, TiO_2, MgO, $SrTiO_3$, $BaTiO_3$, SiO, BeO, CaO, $LiAl_5O_8$, etc.
	Halides	LiF, CaF_2, MgF_2, NaCl, KCl, KBr, CsBr, CsI, AgCl, KRS-5 (TlBr–TlI), KRS-6 (TlBr–TlCl), etc.
	Semiconductors	C (diamond), Si, Ge, Se, Te, etc.
	Chalcogenides	ZnS, CdS, ZnTe, CdTe, ZnSe, PbS, PbSe, PbTe, etc.
	Intermetallic compounds	AlP, AlAs, AlSb, GaP, GaAs, GaSb, InP, InAs, InSb, etc.
Glass	Oxides	SiO_2, GeO_2, GeO_2–Sb_2O_3, TeO_2–ZnO–BaO, etc.
	Halides	$ZnCl_2$, BeF_2, AlF_3, ZrF_4 systems, HfF_4 systems, ThF_4–BaF_2, AlF_3–PbF_3, etc.
	Chalcogenides	As–S, Ge–S, As–Se, Ge–Se, Ge–As–Se, Ge–Sb–Se, Ge–Se–Te, etc.

Table 3.2 Regions of periodic table providing components for infrared materials.

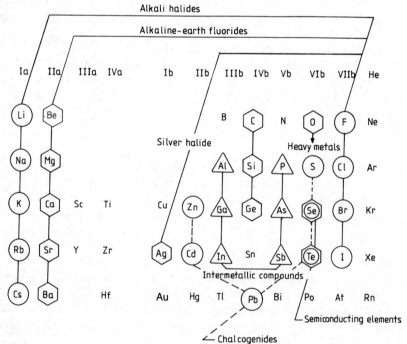

materials. Lines connecting elements show families with similar properties and indicate the formation of important classes of compounds.

These infrared transmitting materials have been reviewed in several books (see, for example, Kruse *et al* (1962), Jamieson *et al* (1963) and Savage (1985)). Optical properties of typical materials have been given by, for example, Hoesterey (1959) and Moses (1971).

3.2.3 Crystalline materials
3.2.3.1 Oxides
The transparent regions of oxide crystals fall below 10 μm, and the regions used widely for infrared optical transmission are restricted to within 4 μm. Typical oxides are SiO_2 (quartz), Al_2O_3 (sapphire), TiO_2 (rutile) and MgO. Of these oxides, SiO_2 is usually used in glass form. Al_2O_3 is extremely hard and thermally stable, and makes excellent window material at wavelengths as long as 6 μm. TiO_2, which is also called titania, is a hard and colorless single crystal which is insoluble in water. MgO is thermally stable and insoluble in water, although the surface upon prolonged exposure to the atmosphere forms a white film that decreases its transmission.

3.2.3.2 Halides
Halide crystals have very wide transparent regions ranging from the ultraviolet to far beyond 10 μm, and so are widely used as infrared optical transmitting windows. The refractive indices of halide crystals are relatively small, resulting in small Fresnel reflections. However, their solubility in water is large compared to other materials such as oxides. Typical halides are LiF, CaF_2, MgF_2, NaCl, KCl, KBr, CsBr, CsI, AgCl, KRS-5, and KRS-6.

LiF and CaF_2 are relatively insoluble in water compared to the other alkali halide crystals. The transparent regions extend to 6 μm and 9 μm for LiF and CaF_2, respectively.

MgF_2 is an excellent infrared transmitting material and has been extensively studied by the Eastman Kodak Company. It is insoluble in water and appears to be resistant to thermal and mechanical shock. The transmission properties of MgF_2 are not affected by temperatures as high as 800 °C.

NaCl, KCl, and KBr are typical alkali halide crystals. Although these crystals are readily available, they are soft and very soluble in water, which means that their use is restricted to environments where they can be carefully protected.

CsBr and CsI are soft and extremely soluble in water. Their transparent regions extend out to as far as 40 μm and 50 μm respectively, although the softness and solubility of these crystals greatly restrict their use in infrared systems.

AgCl is a corrosive, soft, ductile material which is insoluble in water. Unless protected with thin films of silver sulfide, it will darken upon exposure to ultraviolet. The transmitting region extends to 25 μm.

KRS-5 and KRS-6 are polycrystals composed of TlBr–TlI and TlBr–TlCl, respectively. They are slightly soluble in water and toxic. KRS-5 transmits to about 30 μm. As they have a tendency to cold flow, it is important that they are not left unsupported for long periods of time.

3.2.3.3 Semiconducting elements

C (diamond), Si, Ge, Se, and Te can be classified into semiconducting elements. Diamond is transparent at wavelengths above 0.2 μm. Si can transmit the light at wavelengths up to 15 μm except for an absorption at 9 μm. The transparent region of Ge extends to 23 μm. One disadvantage of Si and Ge is that they show large Fresnel reflection losses due to large refractive indices. Another is that transmittance degradation occurs at temperatures above 200 °C, resulting from absorption due to free electrons. Se melts at a low temperature of 490 °C, and so it cannot be used in high temperature regions. Te is an element which transmits light of wavelengths between 3.5 and 8 μm, but is so soft that it must be carefully protected.

3.2.3.4 Chalcogenides

Chalcogenides are compounds composed of chalcogen elements, i.e., S, Se and Te, and elements such as Zn, Cd and Pb. Among the chalcogenides, ZnSe is currently used for the window material transmitting CO_2 laser light. The absorption coefficient for ZnSe at 10.6 μm is of the same order of magnitude as that for KCl, i.e. quite small. Chemical vapor deposition can produce low loss ZnSe polycrystals (P Miles 1976).

3.2.3.5 Intermetallic compounds

Intermetallic compounds are defined here as the compounds between IIIb elements such as Al, Ga and In, and Vb elements such as P, As and Sb. GaAs and InP are well known light emitting materials, and InSb and InAs are well known as materials for infrared light detecting devices. The transparent regions for these compounds reach up to around 10 μm.

The optical and physical properties of the crystalline materials are summarized in table 3.3. It is apparent from the above discussion that there are very many crystalline materials which transmit infrared light, although only a few of them are suitable for use in infrared optical fibers. Heavy-metal oxides and halides such as AgCl and KRS-5 have been studied mainly because of the ease of fiber fabrication, which may

Table 3.3 Properties of typical crystalline materials.

Group	Material	Transparent region (μm)	Refractive index (4 μm)	Young's modulus (10⁶ psi)	Knoop hardness	Density (g cm⁻³)	Melting point (°C)	Water solubility (g/100 g, 20°C)	Thermal expansion coefficient (10⁻⁶ °C⁻¹)	Thermal conductivity (10⁻³ cal s⁻¹ cm⁻¹ °C⁻¹)	Reference
Oxides	SiO_2	0.12–4.5	1.53 ($\parallel$) 1.52 ($\perp$) (2 μm)	14.1–11.1	741 (500 g)	2.65	1743	<0.001	7.97 ($\parallel$) 13.37 ($\perp$)	25.5 ($\parallel$) 14.8 ($\perp$)	a
	Al_2O_3	–5.5	1.73	50–56	1525–2000	3.98	2030	9.8×10^{-5}	6.7 ($\parallel$) 5.0 ($\perp$)	60 ($\parallel$) 55 ($\perp$)	b
	TiO_2	–6.0	2.45 (4.3 μm)	–	880	4.26	1825	Insoluble	9 ($\parallel$) 7 ($\perp$)	28 ($\parallel$) 21 ($\perp$)	b
	MgO	–6.8	1.71	36	690	3.59	2800	1.2×10^{-5}	13 ($\parallel$) 13 ($\perp$)	81 ($\parallel$) 81 ($\perp$)	b
	$LiNbO_3$	0.33–5.2	2.1	–	~5 (Mohs)	4.70	1533	<0.005	16.7	10	a
Halides	LiF	0.11–6	1.35	9.4–11	110	2.6	870	0.27	37	27	b
	CaF_2	0.13–9	1.41	11–15	158	3.2	1360	0.0017	24	23	b
	$NaCl$	0.2–15	1.52	5.8	15.2$\langle 110 \rangle$ 18.2$\langle 100 \rangle$	2.2	801	35.7	44	15	b
	KCl	0.4–21	1.47	4.3	7.2$\langle 110 \rangle$ 9.3$\langle 100 \rangle$	2.0	776	35.4	36	16	b
	KBr	0.21–27	1.54	3.9	5.9$\langle 110 \rangle$ 7.0$\langle 100 \rangle$	2.8	730	53.5	43	11.5	b
	$CsBr$	0.20–40	1.67	2.3	19.5	4.4	636	124.5	48	2.3	b

Table 3.3 (*cont.*)

Group	Material	Transparent region (μm)	Refractive index (4 μm)	Young's modulus (10^6 psi)	Knoop hardness	Density (g cm^{-3})	Melting point (°C)	Water solubility (g/100 g, 20 °C)	Thermal expansion coefficient (10^{-6} °C^{-1})	Thermal conductivity (10^{-3} cal s^{-1} cm^{-1}°C^{-1})	Reference
Halides (cont.)	CsI	0.24–50	1.75	2.3	–	4.5	621	44	50	2.7	b
	AgCl	0.21–25	2.00	2.9	9.5	5.6	458	1.5×10^{-4}	30	2.8	b
	AgBr	0.4–35	2.0 (10.6 μm)	4.64	–	6.47	705	12.6×10^{-6}	34.8	2.9	a
	KRS-5 (TlBr–TlI)	0.5–40	2.38	2.3	30 ⟨110⟩ 38.5⟨100⟩	7.4	1415	0.05	58	1.3	b
	KRS-6 (TlBr–TlCl)	0.21–27	2.19	3.0	40.2	7.2	423	0.32	50	1.7	b
	MgF$_2$	1–8	1.35	16	576	3.2	1396	Insoluble	8	7.5	b
Semiconducting elements	Si	1.2–15	3.43	19.0	1100–1400	2.33	1693	<0.005	4.7	390	b
	Ge	1.8–23	4.02	14.9	700–880	5.33	1209	<0.005	5.5	140	a
	Se (hexagonal)	–	3.7 (∥) 2.9 (⊥)	8.4	2.0 (Mohs)	4.82	490	<0.005	37.9	6	a
	Te (single crystal)	3.5–8.0	6.3 (∥) 4.9 (⊥)	–	18.4	6.25	725	<0.005	−1.6 (∥) 27 (⊥)	15	a

Table 3.3 (cont.)

Group	Material	Transparent region (μm)	Refractive index (4 μm)	Young's modulus (10^6 psi)	Knoop hardness	Density (g cm^{-3})	Melting point (°C)	Water solubility (g/100 g, 20 °C)	Thermal expansion coefficient (10^{-6} °C^{-1})	Thermal conductivity (10^{-3} cal s^{-1} cm^{-1} °C^{-1})	Reference
Chalco-genides	ZnSe (cubic)	0.5–22	2.58 (2 μm)	–	3–4 (Mohs)	5.65	1373	<0.001	7	29	a
Inter-metallic com-pounds	GaAs	1.0–15	3.31	–	750	5.31	1511	<0.005	6.8	108	a
	InP	1.0–14	3.1	–	–	–	1343	–	4.5	–	b
	InSb	7.0–16	4.0 (10 μm)	6.21	–	5.78	523	0.00	4.9	–	c
	InAs	3.8–7.0	3.45	–	380	5.66	1215	<0.05	5.3	64 (poly-crystal)	a

∥ (Electric vector ∥ c-axis)
⊥ (Electric vector ⊥ c-axis)
a Moses (1971)
b Jamieson et al (1963)
c Kruse et al (1962)

result from their relatively low melting temperatures. The other materials are not suitable fiber materials because of, for example, the extremely high melting temperatures in MgF_2 and Al_2O_3.

3.2.4 Glass materials
3.2.4.1 Oxides
The typical oxide glass is the amorphous form of silica, and is sometimes referred to as fused silica or vitreous silica. Its infrared properties are as good as those of crystalline SiO_2, with a transparent region between 0.15 and 4.5 μm. In order to shift the infrared absorption edge toward the longer wavelengths, glass composed of the elements whose atomic numbers are large must be used. Therefore, heavy-metal oxide glasses are preferable to infrared transmitting materials. GeO_2- and TeO_2-based glasses are typical heavy-metal oxide glasses. The transparent region for GeO_2-based glass extends out to beyond 5 μm. The optical and physical properties of those oxide glasses are summarized in table 3.4, and their use in infrared optical fibers is discussed in section 4.1.2.

3.2.4.2 Halides
The most extensively studied halide glasses are fluoride glasses, whose typical compositions are shown in table 4.5. Halide glasses which appeared in the early stages were BeF_2 and then AlF_3. BeF_2 glass has a transparent region between 0.15 and 4.5 μm. However, these glasses both have significant problems—BeF_2 is toxic and hygroscopic, and AlF_3 does not readily form a glass.

$ZnCl_2$ glass is also well known, and is predicted to have a minimum loss of 0.001 dB km^{-1} in the 3.5–4 μm region (Van Uitert and Wemple 1978). Furthermore, the loss of 0.1 dB m^{-1} is also expected at a CO_2 laser light of 10.6 μm in wavelength. However, $ZnCl_2$ glass is extremely hygroscopic, and so it may be difficult to use this glass in usual environments. The detailed nature of this glass will be described in section 4.4.2.

ZrF_4-based glasses are now well known materials and have been studied extensively for use as infrared optical fibers. They contain ZrF_4 as a glass network former, BaF_2 as a primary network modifier, and one or more additional metal fluorides of rare-earths, alkalis, or actinides as glass stabilizers. HfF_4-based glasses show chemical properties similar to ZrF_4-based glasses, but are transparent in a longer wavelength region. These ZrF_4- and HfF_4-based glasses exhibit a lesser tendency toward crystallization and also higher infrared transparency compared to $ZnCl_2$ glasses. Detailed descriptions of the glass properties are given in sections 4.2.2 and 4.2.4. ThF_4–BaF_2 and AlF_3–PbF_3 glasses have also been studied.

HF-based glasses are known as materials which show no toxic or

Table 3.4 Properties of oxide glasses.

Material	Transparent region (μm)	Refractive index	Softening point (°C^{-1})	Thermal expansion coefficient (10^{-7}°C^{-1})	Reference
SiO_2	0.15–4.5	1.458	1670	5.4	Donald and McMillan (1978)
GeO_2		1.62			Olshansky and Scherer (1979)
GeO_2–Sb_2O_3	–5.7	1.72	490	105	Ohsawa et al (1980)
GeO_2–PbO	–5.5	1.91			Donald and McMillan (1978)
CaO–Al_2O_3	0.40–5	1.65	Crystallized	93	Kapany and Simms (1965)
La_2O_3 system	0.40–5	1.79	730	87	Kapany and Simms (1965)
Bi_2O_3–PbO– (BaO, ZnO, Tl_2O)	–7.5	2.5			Donald and McMillan (1978)
As_2O_3–PbO–Bi_2O_3	–5.7				Donald and McMillan (1978)
TeO_2–ZnO–BaO	–6.2	~2.0			Donald and McMillan (1978)
TeO_2–(WO_3/BaO)– (Ta_2O_3, Bi_2O_3/ZnO, PbO)	–6				Boniort et al (1980)

hygroscopic properties (Schröder 1964). Their transparent region is between 0.2 and 5 μm, and they have a comparatively low refractive index of 1.27.

3.2.4.3 Chalcogenides

Chalcogenide glasses are defined as glasses containing at least one of the elements S, Se and Te. As–S and Ge–S glasses are typical chalcogenide glasses, and show a stable vitreous state. The glass forming tendency for ternary chalcogenide glasses increases according to the following order:

S > Se > Te
As > P > Sb
Si > Ge > Sn.

This principle indicates that the glass forming of a Ge–As–S system is much easier than that of Ge–P–S. Also the softening temperature decreases as the atomic numbers of constituent elements increase; that is,

S > Se > Te
P > As > Sb
Si > Ge > Sn.

The infrared absorption edge also shifts toward longer wavelengths as the atomic numbers increase, and so it is necessary to choose the chalcogen elements whose atomic numbers are large for transmitting longer wavelength light, such as that of a CO_2 laser.

Of the chalcogenide glasses, sulfide glasses exhibit no toxicity compared to other chalcogenide glasses such as selenide and telluride glasses. In addition, sulfide glasses have a relatively high softening temperature, as shown in the above discussion, which results in a stable vitreous state. However the infrared absorption edges are restricted to within 10 μm, which means that sulfide glasses cannot be used for CO_2 laser light transmission. Detailed properties of the sulfide glasses will be described in section 4.3.2, together with the other chalcogenide glass materials.

Selenide and telluride glasses, on the other hand, have considerably wider transparent regions. For example, the infrared absorption edge of the ternary Ge–As–Se glass extends out to 16 μm, which is about 6 μm longer than that for Ge–As–S sulfide glass. However, these selenide and telluride glasses have a lower softening temperature, and so attention must be paid to any temperature increase which might occur when these glasses are used for laser power transmission.

The well known selenide glasses are $Ge_{28}Sb_{12}Se_{60}$ (called TI 1173) and $Ge_{33}As_{12}Se_{55}$ (called TI 20), which were developed by Texas Instruments (Hilton *et al* 1975). The properties of selenide and telluride chalcogenide glasses are given in section 4.3.2.

3.3 Classification of infrared optical fibers

As is shown in section 3.1, various infrared optical fibers have been proposed and fabricated to date. In principle, infrared optical fibers can be classified into two groups, i.e., optical fibers and hollow waveguides. The former can be defined as waveguides in which light is guided in solid cores by total internal reflections, while the latter are waveguides whose core regions are hollow, so the light guiding characteristics for each are different. For example, metallic hollow waveguides transmit the light by grazing reflections at the inner surface of the cladding. A detailed classification of infrared optical fibers is shown in table 3.5, together with the minimum losses reported so far.

3.3.1 Optical fibers
3.3.1.1. Glass fibers
Oxide glass fibers. Infrared oxide glass fibers are mainly based on heavy-metal oxides such as GeO_2, GeO_2–Sb_2O_3 and TeO_2. The minimum losses typically occur at wavelengths of around 2–3 μm, and the theoretically predicted loss value is less than 0.1 dB km^{-1}. However, the best value so far reported is 4 dB km^{-1} (GeO_2–Sb_2O_3 glass fiber, Takahashi and Sugimoto 1984, Sugimoto *et al* 1986), which is one order of magnitude larger than that of the conventional silica glass fiber. This discrepancy may result from the existence of the impurities, particularly OH ions.

One of the applications of these fibers is in ultra-low loss transmission lines for long distance optical communication. However, in order to realize such an application, further extensive studies are required, particularly to eliminate the impurities. Alternatively heavy-metal oxide glass fibers can be used as the medium for the non-linear effects. For example, the cross section of Raman scattering of GeO_2 glass fiber is about nine times higher than that of SiO_2 glass fiber. GeO_2 glass fibers can therefore be used as wavelength tunable lasers by combining with high power light sources such as YAG lasers.

Fluoride glass fibers. Fluoride glass fibers are the most promising candidates for the ultra-low loss optical fibers in long distance optical communication. Theoretical prediction shows that fluoride glass fibers possess the lowest transmission losses of the infrared optical glass fibers. From a historical viewpoint, the discovery of stable ZrF_4-based glasses by Lucas and Poulain accelerated the study of fluoride glass (Poulain *et al* 1975), and provided the base for the study of low loss fluoride glass fibers.

Materials for fluoride glass fibers are listed in table 4.5 of section 4.2. As shown in the table, typical materials are ZrF_4- and HfF_4-based glasses, which contain ZrF_4 (or HfF_4) as a glass network former

Table 3.5 Classification of infrared optical fibers.

Classification		Typical examples	Reported minimum loss (dB km^{-1})
Optical fiber	Glass fiber	Oxide glass fiber	GeO$_2$, GeO$_2$–Sb$_2$O$_3$ and TeO$_2$-based glass fibers	4 (1.3 and 2 μm)
		Fluoride glass fiber	ZrF$_4$-based and HfF$_4$-based glass fibers	0.7–0.9 (2–3 μm)
		Chalcogenide glass fiber	As–S, Ge–S, As–Se, Ge–Se, Ge–As–Se, Ge–Sb–Se, and Ge–Se–Te glass fibers	35 (2.44 μm)
	Crystalline fiber	Polycrystalline fiber	TlBr–TlI (KRS-5) and AgCl fibers	70 (10.6 μm)
		Single-crystalline fiber	TlBr–TlI (KRS-5), AgBr, and CsBr fibers	300 (10.6 μm)
Hollow waveguide	Metallic hollow waveguide	Parallel-plate waveguide	Parallel-plate and concave-plate waveguides	200 (10.6 μm)
		Circular waveguide	Circular, helical-circular, and dielectric-coated waveguides	350 (10.6 μm)
	Dielectric hollow waveguide	Dielectric hollow grazing waveguide	SiO$_2$ glass-clad waveguide	–
		Dielectric leaky waveguide		
		Hollow-core fiber	GeO$_2$–ZnO–K$_2$O clad fiber	–
		Liquid-core fiber	C$_2$Cl$_4$ core fiber	20 000 (10.6 μm) ~10^2 (3.39 μm)

(50–70 mol.%), BaF_2 as a primary network modifier (about 30 mol.%), and one or more additional metal fluorides of the rare-earths, alkalis, or actinides as glass stabilizers. Examples are ZrF_4–BaF_2–LaF_3–AlF_3–LiF–PbF_2 and ZrF_4–BaF_2–GdF_3–AlF_3–PbF_2. These glasses exhibit a lesser tendency toward crystallization and higher infrared transparency.

The minimum loss can be obtained at 2–4 μm wavelengths. Losses as low as 0.7–0.9 dB km^{-1} at 2.5 μm have been reported (Tran 1986, Kanamori and Sakaguchi 1986). This may be the lowest value reported so far in infrared optical fibers, making fluoride glass fibers the best candidates for ultra-low loss optical fibers.

In addition, fluoride glasses have the advantage of providing a great deal of compositional flexibility, which allows them to be tailored to a broad range of properties essential for forming compatible core and cladding materials. Furthermore, material dispersion plays an important role when these fibers are used as transmission lines for long distance optical communications. At the wavelength where the material dispersion is zero, the light pulses can be transmitted through the optical fiber without any distortion, which leads to a considerable increase in information transmission capacity. The zero material dispersion wavelength is less than 2 μm for almost all fluoride glass fibers. This value is slightly different from the wavelength giving the minimum loss, but the magnitude of the material dispersion is small enough through an extended range of wavelengths, including the minimum loss region. Moreover, it is possible to exploit waveguide dispersion in fibers to compensate for material dispersion. This makes it possible to reduce the dispersion at the wavelength giving the minimum loss. This feature of fluoride glass fibers is another reason why they are suitable for use as ultra-low loss optical fibers.

Chalcogenide glass fibers. Glasses containing at least one of the chalcogen elements, i.e., S, Se, or Te, are generally called chalcogenide glasses. Materials used for chalcogenide glass fibers are shown in tables 4.12, 4.13 and 4.14 of section 4.3.

Sulfide glass fibers can transmit light of wavelength 2–5 μm. The transmission loss has been reduced to less than 0.1 dB m^{-1} (Kanamori *et al* 1985). However, further loss reduction is thought to be difficult because of the existence of an intrinsic weak absorption tail. Since the sulfide glass fibers are thermally stable, they can be used for the short distance transmission such as infrared image transmission (by a bundle fiber) and CO laser (wavelength 5.3 μm) light transmission. These applications are important in the fields of infrared sensing systems, and laser surgery and machining.

On the other hand, selenide and telluride glass fibers have a wide transparency range, which reaches up to around 10 μm in wavelength.

However, the predicted minimum transmission losses are higher than those of sulfide glass fibers because of a lesser tendency toward vitrification. Therefore the main target of the research into these chalcogenide glass fibers is CO_2 laser power transmission at a 10.6 μm wavelength. Although the transmission loss of about 1 dB m^{-1} has been already achieved (Katsuyama and Matsumura 1986), further loss reduction is still needed for stable power transmission. In contrast, since wide bandwidth infrared light transmission is possible with Ge–Se or As–Se based glass fibers, radiometric temperature measurement using these glass fibers has been realized.

Other glass fibers. $ZnCl_2$ glass fibers were the first to be proposed as candidates with a potential for ultra-low loss (Van Uitert and Wemple 1978). However, it was found that $ZnCl_2$ is disadvantageous because it shows hygroscopic behaviour which makes fiber fabrication difficult. Therefore, at present, $ZnCl_2$ glass fibers are not being studied extensively. $ZnBr_2$, AgI–AgF–AlF$_3$, and $PbBr_2$–PbI_2 glasses have also been studied as the materials for infrared optical fibers.

3.3.1.2 Crystalline fibers
Polycrystalline fibers. Typical materials so far studied are TlBr–TlI (which is usually called KRS-5), AgCl, and AgCl–AgBr, which are characterized by a relatively low melting temperature and a high tensile strength. Table 5.1 in section 5.1 shows materials used for polycrystalline fibers. In general, polycrystalline fibers show low losses in the wavelength region of more than 10 μm. This is useful for transmitting CO_2 laser power. A loss of as low as 0.1 dB m^{-1} at 10.6 μm wavelength was achieved in various research institutes (see table 5.1). Furthermore, the CO_2 laser power exceeding 130 W could be transmitted through the polycrystalline fiber. CO_2 laser power transmission can be applied in laser surgery and laser welding.

It should be noted, however, that a new problem occurred for the loss property—loss increase due to plastic deformation and grain growth of the fiber material. In addition, water in the atmosphere is found to increase the transmission loss because of the corrosion of halides caused by the water.

Single-crystalline fibers. Materials for single-crystalline fibers are almost the same as those for polycrystalline fibers. TlBr-TlI, AgBr, KCl, CsBr, and CsI were mainly studied and various fabrication methods based on the crystal growth techniques were proposed. The materials used for the single-crystalline fibers can be seen in table 5.2 of section 5.2. The transmission loss of 0.3 dB m^{-1} and maximum transmission power of 47 W were obtained at a 10.6 μm wavelength by using a 1 mm diameter CsBr fiber (Mimura and Ota 1982).

The advantage of the single-crystalline fibers is that they possess a wide transparent wavelength region from visible to far-infrared. This makes it possible to transmit both visible and infrared light, so that these fibers can be used in laser scalpels with a guiding function using visible light. On the other hand, the disadvantage is that there is significant loss increase due to plastic deformation caused by repeated bending. Therefore these fibers must be handled carefully, for example, by introducing a method of preventing a high bending curvature.

Single-crystalline fibers made of Al_2O_3 and Ge are also important in transmitting infrared light. These fibers can be used for a variety of fields such as micro-laser rods and thin semiconductor elements.

3.3.2 Hollow waveguides
3.3.2.1 Metallic hollow waveguides
Parallel-plate waveguides. Parallel-plate waveguides are constructed, for example, from a pair of aluminum sheet strips which form the top and bottom walls, and a pair of brass shimstocks which form the sidewalls. The light guiding of these waveguides is based on the grazing reflection at the inner surface of the metallic wall. The grazing reflection can be greatly enhanced by reducing the incident angle of the light. More than 200 W c.w. CO_2 laser power has been transmitted through parallel-plate waveguides, representing 80% overall transmittance (Garmire *et al* 1979). Various modifications of the parallel-plate waveguide have been proposed so far. For example, concave-plate waveguides were proposed so as to confine the transmitted light to the central region.

The advantage of these waveguides is that they can be relatively easily fabricated compared to infrared optical fibers. Furthermore, the hollow core is effective in dissipating heat during power transmission. Laser scalpels using parallel-plate metallic waveguides have been constructed which can transmit CO_2 laser power of 95 W (Kubo and Hashishin 1986). However, these waveguides are not as flexible as optical fibers, and so applications are limited.

Circular waveguides. Circular metallic waveguides show a low loss for TE mode operation, whereas the hybrid HE modes, which can be excited efficiently by commercially available lasers, give a high transmission loss. On the other hand, a dielectric-coated circular hollow metallic waveguide has a low loss of the HE mode (Miyagi *et al* 1983). A nickel pipe waveguide with a germanium inner coating gives a low loss of 0.35 dB m^{-1} at a 10.6 μm wavelength. In addition, bending loss can be reduced by coating with a dielectric layer.

Furthermore, helical-circular waveguides were proposed which are based on the whispering-gallery principle, i.e., grazing reflections only at a metallic surface on one side (Marhic *et al* 1978). This waveguide is basically expected to give a low loss. However, experiments on the transmission properties using a CO_2 laser have not yet been reported.

3.3.2.2 Dielectric hollow waveguides

Dielectric hollow grazing waveguides. The light guiding by grazing reflection can be applied to not only metallic waveguides but also dielectric waveguides. Since these waveguides have, however, a relatively high transmission loss and require very precise control of the cladding surface, high quality waveguides have not yet been fabricated.

Dielectric leaky waveguides. Dielectric leaky waveguides can transmit the light by using leaky modes whose losses are basically low (Miyagi and Nishida 1980). Leaky waveguides can be characterized by their thin tube layers, while the conventional dielectric hollow waveguides can be defined as having thick dielectric layers whose outer surfaces never affect the transmission properties. These waveguides are particularly advantageous because they can transmit the hybrid HE_{11} mode with very low loss. This is preferable for the connection with the laser because commercially available lasers have the same light field pattern as the hybrid HE_{11} mode light.

Applications of the leaky waveguides tend to be in CO_2 laser power transmission over a short distance. However, since fabrication is very difficult only a few experimental results have yet been reported.

Hollow-core fibers. Hollow-core optical fibers, as proposed by Hidaka *et al* (1982), are rather different from the other hollow waveguides. These fibers transmit infrared light by total reflection between the hollow core and the dielectric cladding. This light guiding is possible only in the abnormal dispersion region where the refractive index is lower than unity.

GeO_2–ZnO–K_2O was found to be a suitable material for an infrared transmitting hollow-core fiber (Hidaka *et al* 1982). The minimum transmission loss obtained by the experiment was $20 \, dB \, m^{-1}$, although the predicted loss is less than $0.1 \, dB \, m^{-1}$ at a $10.6 \, \mu m$ wavelength. If a smooth inner surface is formed, the transmission loss may be reduced further.

Liquid-core fibers. Liquid-core fibers contain liquid materials in the hollow cores. These fibers are advantageous because there are no stress effects leading to birefringence, and wall imperfections and scattering effects are negligible. The transmission losses and the transparent wavelength regions depend on the liquids used. A C_2Cl_4-filled fused-quartz fiber has been fabricated, but the transmission loss remained very high ($10^4 \, dB \, km^{-1}$) because of the high concentration of impurities.

3.4 Applications of infrared optical fibers

3.4.1 Introduction
Infrared optical fibers are applicable to both long distance optical communications and short haul light transmissions. In long distance optical communications, ultra-low loss as well as low dispersion properties are required. Short haul applications involve nuclear radiation-resistant optical transmission, infrared remote sensing such as temperature measurement by thermal radiation, and laser power transmission such as laser surgery.

3.4.2 Long distance optical communications
Since Pinnow *et al* (1978), Van Uitert and Wemple (1978) and Goodman (1978) discussed the possibility of ultra-low loss reaching less than $0.01 \, dB \, km^{-1}$, research into non-silica-based infrared fibers has become one center of interest in the optical fiber field. These ultra-low loss optical fibers can be used in long distance optical communication. In particular, nonrepeated intercontinental undersea optical communication is the most attractive application, because the absence of repeaters significantly increases the reliability of communication.

Single-mode optical fibers are essential for long distance optical communication, since they have a wide bandwidth. The maximum bandwidth usually appears around the wavelength where the material dispersion falls to zero. Fortunately, the wavelength giving the minimum loss nearly coincides with the zero material wavelength for almost all infrared optical fibers.

The achievement of ultra-low loss depends strongly on the materials used. Requirements for the materials are as follows.

(i) The tail of the lattice absorption edge in the infrared must be at a wavelength sufficiently far from that of interest.

(ii) Scattering losses such as Rayleigh scattering loss must be minimized.

(iii) Impurity and defect absorptions must be small, i.e., $< 10^{-9} \, cm^{-1}$.

Materials satisfying the above requirements include heavy-metal oxide glass, halide crystals, halide glass, and chalcogenide glass. Of these the potential for low loss has been shown only by a fluoride glass that is classified among the halide glasses, although the reported loss is still around $1 \, dB \, km^{-1}$, which is one order of magnitude larger than that of the silica optical fibers currently used. The minimum losses of other fibers so far reported are $4 \, dB \, km^{-1}$ at $1.3 \, \mu m$ for GeO_2 heavy-metal oxide fiber (Sugimoto *et al* 1986), $70 \, dB \, km^{-1}$ at $10.6 \, \mu m$ for AgBr polycrystalline fiber (Takahashi *et al* 1986), and $35 \, dB \, km^{-1}$ at $2.44 \, \mu m$ for As–S chalcogenide glass fiber (Kanamori *et al* 1984).

It can be concluded that the most appropriate material for the ultra-low loss optical fiber is fluoride glass. However, it should be stressed that the research on ultra-low loss optical fibers is still under way.

3.4.3 Nuclear radiation-resistant optical fibers
It is very important for optical fibers to be able to transmit light even under nuclear radiation. If radiation-resistant optical fibers were available, light transmission in nuclear power plants would become possible, resulting in the stable control of atomic power. Furthermore, applications to some military fields may be possible.

In general, the effect of ionizing radiation on optical fibers is to cause a considerable loss increase due to color center formation. Defects in the materials are responsible for the color centers whose absorptions usually take place in the ultraviolet region. This fact is especially relevant to those infrared optical fibers whose minimum transmission losses appear in wavelengths far from the ultraviolet.

Radiation effects on the light transmission properties have been studied mainly for ZrF_4-based fluoride glass fibers. This is due to the fact that fluoride glass fibers possess the lowest loss properties. Experiments show that the radiation-induced losses in the ZrF_4-based fluoride glass fibers are 1000–2000 dB km^{-1} in the near-infrared regions when the dose is 10^6 roentgen (Ohishi *et al* 1983). One may feel that this value is comparable to those of conventional silica glass fibers whose absorption bands due to color centers are close to the light transmission wavelength. However, if the purity of the fiber material is improved further, the loss increase by radiation may be reduced to some extent. This is due to the fact that some impurities are responsible for the color center formation. It is interesting, however, that some dopants act as inhibitors to color center formation. For example, Fe- and Ti-doped fluoride glasses become considerably radiation harder (Ohishi *et al* 1985).

The research on the radiation-resistant infrared optical fibers is in its early stages at present. However, by optimizing the compositions of the fiber materials, further improvements to the radiation-resistant properties are expected in the near future.

3.4.4 Infrared remote sensing
Applications of infrared remote sensing include radiometric temperature measurements, infrared image transmissions and remote sensing of infrared spectroscopy. When infrared optical fibers are used, the temperature, infrared image and infrared spectra may be easily obtained even in unfavorable surroundings such as in engines, furnaces and

nuclear reactors. In particular, the use of infrared optical fibers is advantageous because they can transmit longer wavelength light that makes it possible, for instance, to measure lower temperatures such as room temperature.

Radiometric temperature measurements are carried out by detecting infrared light radiated from heated bodies. Since the spectra of the radiated light can be described approximately by Planck's formula, the longer wavelength light must be observed in order to estimate the lower temperature. For example, the room temperature corresponds to the wavelength region around 8–10 μm. Experiments performed so far show that both TlBr–TlI crystalline fibers (Shimizu and Kimura 1986) and Te-based chalcogenide glass fibers (Katsuyama *et al* 1985) can measure temperatures of around 50 °C. Besides those experiments, many reports on the temperature measurements have been published to date.

Infrared image transmissions can also be done by using infrared bundle fibers. This method is useful for mapping the temperature distribution of an object far from the observer. Furthermore, it is useful for measuring the temperature distribution of the human body, particularly of internal organs such as the stomach, because in this way the existence of abnormalities such as cancer may be detected. However, in order to succeed in measuring such subtle discrepancies of the temperature, further loss reduction for longer wavelength is needed.

3.4.5 *Laser power transmissions*

Laser power deliveries through optical fibers are one of the most interesting research targets in infrared fiber studies. The first experiment on laser power transmission was performed by Pinnow *et al* (1978). They announced a 2 W continuous CO_2 laser beam transmission through a polycrystalline fiber. Since then, various optical fibers have been studied for the laser power transmission.

The main application of laser power transmission is in laser surgery. CO and CO_2 lasers (whose emitting wavelengths are 5.3 and 10.6 μm, respectively) are used as scalpels. The laser power is required to be around 100 W for such applications. Laser welding and machining are also valuable applications; however, the required laser power is relatively high compared to the laser surgery, and so application is more difficult.

Infrared optical fibers that can be used in power transmission are chalcogenide glass, polycrystalline and single-crystalline fibers. For example, laser power of more than 50 W at a 10.6 μm wavelength has been transmitted through TlBr–TlI and Ag halide polycrystalline fibers. In addition, it has been reported that sulfide glass fibers can transmit laser power of 40 W at CO laser wavelength (5.3 μm). Although the hollow waveguides exhibit less flexibility, they can transmit laser power of up to 200 W (10.6 μm). The detailed values of transmitted laser powers are described in section 7.4.2.

There are many reports describing the experimental results of various applications of laser power transmission. Laser scalpels have been widely studied. The fiber materials mainly used for laser scalpels are TlBr–TlI polycrystals which are fabricated by an extrusion method, and transmit laser power of as high as 130 W. The use of chalcogenide glass fibers and parallel-plate metallic waveguides in laser scalpels has also been reported.

3.4.6 Other applications

Besides the applications described in the previous sections, there are many other applications of infrared optical fibers. For instance, GeO-based oxide glass fibers can be applied to devices using non-linear optical effects, namely fiber Raman lasers that can be used as tunable laser light sources. Since the cross section of Raman scattering of GeO_2 glass is about nine times higher than that of SiO_2 glass, the higher order Raman lines, that reach up to ninth order, can be easily obtained (Sugimoto *et al* 1986). These light sources are expected to be applied to various spectroscopic analyses.

Furthermore, magneto-optical effects of infrared optical fibers can be used in magneto-optical switches, power-controllers, and other sensing elements for magnetic fields in the infrared region. TlBr–TlI polycrystalline fibers and As–S chalcogenide glass fibers have been examined in order to verify the magneto-optical effects (Sato *et al* 1983, 1985). The measured values of Verdet constants, which characterize the efficiency of the magneto-optical effect, are 5×10^{-4} deg Oe^{-1} cm^{-1} for TlBr–TlI polycrystalline fiber and 1.62×10^{-2} min cm^{-1} G^{-1} for As–S chalcogenide glass fiber.

3.4.7 Summary

Applications of infrared optical fibers are now being studied extensively in various research institutes. The main targets are for long distance optical communications using ultra-low loss infrared fibers and for laser power transmissions at CO and CO_2 laser wavelengths. The former is useful for undersea intercontinental communications and the latter may be widely used for laser surgery and laser welding. Many other applications are possible, such as infrared remote sensing.

However, it should be noted that further loss reduction is still required for achieving such applications, particularly applications for long distance optical communications.

3.5 Measurement techniques for transmission properties

3.5.1 Introduction

In order to evaluate the infrared optical fibers, various properties such as transmission losses and refractive indices in the infrared region must be measured. Measurements in the infrared region are generally more difficult than those in the visible region because of the unstable measurement instruments. In this section, transmission loss measurements are described for infrared bulk samples and infrared optical fibers, and refractive index and dispersion measurements are also discussed.

3.5.2 Loss measurements for infrared bulk samples

It is very important to estimate the absorption coefficients (which are equivalent to the transmission losses) of infrared materials before fiber drawing. Intrinsic losses, such as multi-phonon absorptions, can easily be estimated from measurements on bulk samples. However, since the absorption coefficients of materials for infrared optical fibers are extremely small, special techniques are needed to evaluate the losses precisely. Various loss measurement techniques have been proposed so far, the most important of which are differential spectrophotometry and thermometric laser calorimetry.

The differential spectrophotometry technique (Deutsch 1973) is based on the measurement of the absorption difference between samples with the same qualities but different thicknesses. In this set-up, the reflections at the surfaces cancel out. Hidaka *et al* (1980) improved the accuracy by introducing a precisely controlled slit and a data-averaging program. A resolution better than 0.1% was obtained in the 300–1400 cm^{-1} frequency region, which includes the CO_2 laser frequency.

The thermometric laser calorimetry technique is based on the measurement of the temperature rise of a sample irradiated with a beam of known power. An experimental arrangement for laser calorimetry is shown in figure 3.2. In this method, a sample is thermally isolated from its surroundings and allowed to come to thermal equilibrium. Surface absorption is sometimes the origin of measuring error. An absorption coefficient as low as 7×10^{-6} cm^{-1} can be measured using this method (Hass *et al* 1975). It has also been reported (Nister *et al* 1984) that by using miniature thermistors as temperature sensors the sensitivity can be at least two orders of magnitude higher than in thermocouple calorimetry.

A photoacoustic technique can be used for laser calorimetry instead of temperature change measurement. When a periodically interrupted light beam is absorbed in a medium, an acoustic or elastic wave is generated by the light-induced heating and subsequent expansion. Hordvik and

Schlossberg (1977) improved the technique by placing a strain transducer directly in contact with the sample. The sensitivity of this technique was found to be limited to about $1 \times 10^{-5}\,\mathrm{cm}^{-1}$ due to radiation scattered onto the transducer, but the technique is predicted to be capable of measuring absorption coefficients in the $10^{-6}\,\mathrm{cm}^{-1}$ range using laser powers of about 1 W.

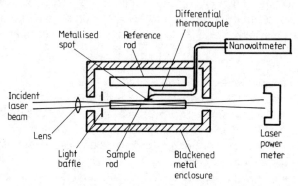

Figure 3.2 The set-up for laser calorimetry (after Hass *et al* 1975).

Furthermore, interferometric calorimetry has been proposed for measuring the absorption coefficient which is hardly influenced by the surface conditions of the sample (Itoh and Ogura 1982). In this method, the temperature rise caused by the absorption of a laser beam is calculated by the change of optical path length of the sample, which arises because the refractive index and sample thickness depend on temperature. Michelson interferometry and holographic interferometry are used to measure the change of optical path length; holographic interferometry is particularly advantageous because it is not restricted by sample shape. The advantage of interferometric calorimetry is that it can separate surface absorption from bulk absorption, so that an absorption coefficient can be obtained which is hardly influenced by surface condition. A variety of modifications of the laser calorimetry have been proposed in order to improve measurement accuracy.

3.5.3 Loss measurements for infrared optical fibers
The basic method of spectral loss measurement is similar to that for conventional silica glass fibers, and is based on the measurement of the difference between the transmittance of a long fiber and that of a cut-back short fiber. The measurement in the infrared region requires a number of detectors because the spectral response of an individual detector cannot cover the whole infrared spectral region.

Spectral loss measurement has been carried out in a wide wavelength range from visible to middle infrared by changing detectors (Jinguji *et al* 1982a). The 0.7–1.8 µm band can be covered with the conventional test set developed for the silica fiber; for example, a tungsten lamp and photomultiplier are used as the light source and detector. In the wavelength range from 1.8 to 11 µm, the optoelectronic system is specially designed, and is shown in figure 3.3. This system is composed of a platinum resistance heater as a light source, a step scanning monochromator, an InSb detector or a HgCdTe detector cooled with liquid N_2, a current-sensitive preamplifier, and a hard copy unit. The platinum lamp with 100 W power generates high output power in the infrared. In order to eliminate the second-order spectrum in the output spectrum from the grating monochromator, infrared bandpass filters are used. The optical system is composed of infrared mirrors to avoid measurement errors arising from the fact that the numerical aperture changes with wavelength in a system using a glass or crystal lens. The numerical aperture in the light injection can be varied in the range from 0.01 to 0.2.

The fiber output signals are detected with InSb in the 1.8–5.5 µm and with HgCdTe in the 5–11 µm wavelength regions. The detected signals are amplified by the current-sensitive preamplifier and fed into the lock-in amplifier. The amplified signals are measured with a signal voltmeter and then displayed by the minicomputer. Alternative light sources and detectors can be used instead of a platinum lamp and HgCdTe detector. For example, a nichrome wire or SiC is a stable light source in the infrared region, and a thermocouple is also a stable detector covering a wide spectrum region.

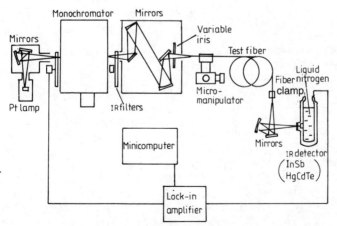

Figure 3.3 The spectral loss measurement system for infrared optical fibers (after Miyashita and Manabe 1982, © 1982 IEEE).

In addition to spectral loss, transmission losses can be measured at several specific wavelengths by using various lasers such as HF, DF, CO and CO_2 lasers.

3.5.4 Refractive index measurements

The refractive indices of bulk materials before fiber drawing are usually measured by a minimum deviation method. A triangular prism made from the infrared bulk sample is used, and the angle of the refracted light beam through the prism is precisely measured. Jinguji *et al* (1982b) reported the minimum deviation method in the infrared region. A precision spectrometer (Kalnew GMR-1S) is used to determine the spectral refractive index. It covers the wavelength range 0.4–4.26 μm. Refractive indices are measured at 42 wavelengths. Mercury, helium, and hydrogen emission lines are used for wavelengths up to 2.5 μm. Beyond 2 μm, a platinum wire heater is used and the measuring wavelengths are identified by the absorption bands of 1, 2, 4-trichlorobenzene and polystyrene. A photomultiplier with high sensitivity is used as a detector for the emission line spectra at 0.4–0.8 μm, and an indium antimonide detector operated at liquid nitrogen temperature is used for the longer wavelengths. In both cases, the incident light flux is chopped at 270 Hz and the detected optical signal is amplified by a low noise current sensitive preamplifier. The amplified output signal is fed into a lock-in amplifier and then is recorded as a function of the minimum deviation angles for each line. All measurements are made at $23 \pm 1.0\,°C$. Measurement error of the refractive index is within 0.0001.

The measured refractive indices are fitted to the conventional dispersion of the following form:

$$n(\lambda) = \frac{A}{\lambda^4} + \frac{B}{\lambda^2} + C + D\lambda^2 + E\lambda^4, \qquad (3.1)$$

where λ is the wavelength and coefficients A–E are constants. Material dispersion $M(\lambda)$ can be obtained as

$$M(\lambda) = -\frac{\lambda}{c}\frac{d^2n(\lambda)}{d\lambda^2} = \frac{2}{c}\left(\frac{10A}{\lambda^5} + \frac{3B}{\lambda^3} + D\lambda + 6E\lambda^3\right) \qquad (3.2)$$

where c is the speed of light.

References

Boniort J Y, Brehm C, Dupont P H, Guignot D and LeSergent C 1980 *Technical Digest of 6th European Conf. on Optical Communication* 61–4
Deutsch T F 1973 *J. Phys. Chem. Solids* **34** 2091–104
Donald I W and McMillan P W 1978 *J. Mater. Sci.* **13** 2301–12
Garmire E, McMahon T and Bass M 1976 *Appl. Phys. Lett.* **29** 254–6
—— 1979 *Appl. Phys. Lett.* **34** 35–7

Goodman C H L 1978 *Solid-State Electron Dev.* **2** 129–37
Harrison W B 1975 *Honeywell Inc. Final Tech. Rep.* (July) AFML-TR-75-104
Hass M, Davisson J W, Rosenstock H B and Babiskin J 1975 *Appl. Opt.* **14** 1128–30
Hettner G and Leisegang G 1948 *Optik* **3** 305–14
Hidaka T, Kumada K, Shimada J and Morikawa T 1982 *J. Appl. Phys.* **53** 5484–90
Hidaka T, Morikawa T and Shimada J 1980 *Appl. Opt.* **19** 3763–6
Hilton A R, Hayes D J and Rechtin M D 1975 *J. Non-Cryst. Solids* **17** 319–38
Hilton A R, Jones C E, and Brau M 1966 *Infrared Phys.* **6** 183–94
Hoesterey H F 1959 Infrared transmitting materials *Electronics* Jan.
Hordvik A and Schlossberg H 1977 *Appl. Opt.* **16** 101–7
Itoh M and Ogura I 1982 *J. Appl. Phys.* **53** 5140–5
Jamieson J A, McFee R H, Plass G N, Grube R H and Richards R G 1963 *Infrared Physics and Engineering* (New York: McGraw-Hill)
Jinguji K, Horiguchi M and Manabe T 1982a *Appl. Opt.* **21** 571–2
Jinguji K, Horiguchi M, Shibata S, Kanamori T, Mitachi S and Manabe T 1982b *Electron. Lett.* **18** 164–5
Joos G 1948 *F.I.A.T. Review of German Science, The Physics of Solids*; Part II 1934–46
Kanamori T and Sakaguchi S 1986 *Japan. J. Appl. Phys.* **25** L468–70
Kanamori T, Terunuma Y, Takahashi S and Miyashita T 1984 *J. Lightwave Technol.* **LT-2** 607–13
—— 1985 *J. Non-Cryst. Solids* **69** 231–42
Kapany N S and Simms R J 1965 *Infrared Phys.* **5** 69–80
Kapron F P, Keck D B and Maurer R D 1970 *Appl. Phys. Lett.* **17** 423–5
Katsuyama T and Matsumura H 1986 *Appl. Phys. Lett.* **49** 22–3
Katsuyama T, Matsumura H and Kawakami H 1985 *Technical Digest of Annual Meeting of the Institute of Electronics and Communication Engineers of Japan* 1037 (in Japanese)
Kruse P W, McGlauchlin L D and McQuistain R B 1962 *Elements of Infrared Technology: Generation, Transmission, and Detection* (New York: John Wiley)
Kubo U and Hashishin Y 1986 *Technical Digest of 22nd Symposium in Institute of Electrical Communication, Tohoku University, Sendai, Japan* 43–52 (in Japanese)
Land C E, Thatcher P D and Haertling G H 1974 *Applied Solid State Science* Vol 4, ed R Wolfe (New York: Academic Press) pp 137–233
Marhic M E, Kwan L I and Epstein M 1978 *Appl. Phys. Lett.* **33** 609–11
Melloni M 1833a *Ann. Chim.* **53** 5–73
—— 1833b *Ann. Chim.* **55** 337–97
Miles P 1976 *Opt. Engng* **15** 451–9
Miles P A 1976 *Proc. 5th Annual Conf. on IR Laser Window Materials* 7–15
Mimura Y and Ota C 1982 *Appl. Phys. Lett.* **40** 773–5
Miya T, Terunuma Y, Hosaka T and Miyashita T 1979 *Electron. Lett.* **15** 106–8
Miyagi M, Hongo A, Aizawa Y and Kawakami S 1983 *Appl. Phys. Lett.* **43** 430–2
Miyagi M and Nishida S 1980 *IEEE Trans. Microwave Theory Tech.* **MTT-28** 398–401

Miyashita T and Manabe T 1982 *IEEE J. Quantum Electron.* **QE-18** 1432–50
Moses A J 1971 *Optical Materials Properties, Handbook of Electronic Materials* Vol. 1 (New York: IFI/Plenum)
Nishihara H, Inoue T and Koyama J 1974 *Appl. Phys. Lett.* **25** 391–3
Nister L C, Nister S V and Teodorescu V 1984 *J. Appl. Phys.* **56** 6–9
Ohishi Y, Mitachi S, Takahashi S and Miyashita T 1983 *Electron. Lett.* **19** 830–1
—— 1985 *IEE Proceedings* **132** 114–8
Ohsawa K, Takahashi H, Shibata T and Nakamura K 1980 *Technical Digest of Annual Meeting of the Institute of Electronics and Communication Engineers of Japan* pp 4–189 (in Japanese)
Olshansky R and Scherer G W 1979 *Proc. 5th ECOC and 2nd IOOC Amsterdam* 12.1.1–12.5.3
Parsons W F 1972 *Appl. Opt.* **11** 43–9
Pinnow D A, Gentile A L, Standlee A G, Timper A J and Hobrock L M 1978 *Appl. Phys. Lett.* **33** 28–9
Poulain M, Poulain M, Lucas J and Brun P 1975 *Mater. Res. Bull.* **10** 243–6
Sakaguchi S and Takahashi S 1986 *Tech. Dig. 22nd Symposium in Institute of Electrical Communication, Tohoku University, Sendai, Japan* 1–9 (in Japanese)
Sakuragi S, Saito M, Kubo Y, Imagawa K, Kotani H, Morikawa T and Shimada J 1981 *Opt. Lett.* **6** 629–31
Sato H, Kawase M and Saito M 1985 *Appl. Opt.* **24** 2300–3
Sato H, Tsuchiya E and Sakuragi S 1983 *Opt. Lett.* **8** 180–2
Savage J A 1985 *Infrared Optical Materials and their Antireflection Coatings* (Bristol: Adam Hilger)
Schröder J 1964 *Angew. Chem. inter. Edit.* **3** 376
Shimizu M and Kimura M 1986 *Technical Digest of 22nd Symposium in Institute of Electrical Communication, Tohoku University, Sendai, Japan* 142–50 (in Japanese)
Stöber F 1925 *Z. Krist.* **61** 299–314
Strong J 1930 *Phys. Rev.* **36** 1663–6
Sugimoto I, Shibuya S, Takahashi H, Kachi S, Kimura M and Yoshida S 1986 *Technical Digest of 22nd Symposium in Institute of Electrical Communication, Tohoku University, Sendai, Japan* 10–21 (in Japanese)
Takahashi H and Sugimoto I 1984 *J. Lightwave Technol.* **LT-2** 613–5
Takahashi K, Yoshida N and Yamauchi K 1986 *Technical Digest of 22nd Symposium in Institute of Electrical Communication, Tohoku University, Sendai, Japan* 30–5 (in Japanese)
Tran D C 1986 presented at *Conf. on Optical Fiber Communication (Atlanta)*
Van Uitert L G and Wemple S H 1978 *Appl. Phys. Lett.* **33** 57–9
Yoshida S 1986 *Kougaku gijutsu contact* **24** 681–91 (in Japanese)

4 Glass Fibers for Infrared Transmission

Glass fibers are the most promising candidates for infrared optical fibers, and can be made from a variety of materials, such as oxides, halides and chalcogenides. In this chapter oxide, fluoride and chalcogenide glasses are described.

4.1 Oxide glass fibers

4.1.1 Introduction

SiO_2-based glass fibers are widely used for optical communication systems and also for optical measurement systems, because the transparency and mechanical properties are superior to those of other materials. However, new materials whose transmission losses are expected to be lower than the SiO_2-based glass have been proposed recently. Heavy-metal oxide glasses, such as GeO_2 glass, are thought to be good candidates for ultra-low loss optical fibers. Olshansky and Scherer (1979) calculated that a transmission loss of less than $0.1 \, dB \, km^{-1}$ could be achieved around 2.2–2.4 μm. Another advantage of the heavy-metal oxide glass fibers is that almost the same fabrication method as that of the SiO_2-based glass fibers can be used, so that there is basically no difficulty in fabricating the fibers.

It should be noted that the heavy-metal oxide glass fibers such as GeO_2 glass fibers have a large non-linear effect compared to the SiO_2-based glass fibers. Cross sections of the Raman scattering for the GeO_2 glass optical fibers are one order of magnitude larger than those of the SiO_2-based glass fibers. GeO_2 glass optical fibers therefore have a potential for application to the new fields of non-linear optics, such as fiber Raman lasers which can be used as wavelength-tunable lasers.

4.1.2 Materials

In 1965 Kapany and Simms (1965) fabricated lanthanate glass optical fibers; this may be the first heavy-metal oxide glass fiber actually made. However, details of the fiber fabrication and its optical properties were not given. Recently, germanate glass was suggested as useful for infrared optical fiber with a minimum loss of less than $0.2 \, \text{dB km}^{-1}$ at a $2.5 \, \mu\text{m}$ wavelength and a zero material dispersion wavelength of near $1.69 \, \mu\text{m}$ (Olshansky and Scherer 1979).

The advantage of heavy-metal glasses such as GeO_2 glass is that since their constituent metals (such as Ge) are heavier than Si in SiO_2 glass, infrared absorption due to lattice vibration (Ge–O) can be shifted toward a longer wavelength. This leads to the ultra-low loss in the infrared region. The heavy-metal oxide glasses so far proposed are listed in table 4.1. In principle, these glasses can be divided into GeO_2-based and TeO_2-based glasses, as shown in the table. Figures 4.1 and 4.2 show the transmission properties of GeO_2-based glasses (Osawa *et al* 1980) and TeO_2-based glasses (Boniort *et al* 1980), respectively. It can be seen from the figures that the intrinsic absorptions due to lattice vibrations are shifted to the wavelengths above $5 \, \mu\text{m}$.

Table 4.1 Heavy-metal oxide glasses proposed for infrared optical fibers.

Material	Structure	Fabrication	Property	Reference
Lanthanate	–	–	–	Kapany and Simms (1965)
GeO_2	–	–	$0.15 \, \text{dB km}^{-1}$ ($1.8 \, \mu\text{m}$, predicted value)	Olshansky and Scherer (1979)
GeO_2	GeO_2–SiO_2 cladding	VAD	$4 \, \text{dB km}^{-1}$ ($1.3 \, \mu\text{m}$)	Sugimoto *et al* (1986)
GeO_2–Sb_2O_3	GeO_2 cladding	VAD	$4 \, \text{dB km}^{-1}$ ($2 \, \mu\text{m}$)	Takahashi and Sugimoto (1984)
GeO–($BiO_{1.5}$–$TlO_{0.5}$–PbO–$SbO_{1.5}$)	–	Crucible	–	Wood *et al* (1982)
TeO_2–BaO–ZnO	–	Crucible	$1 \, \text{dB m}^{-1}$ ($2 \, \mu\text{m}$)	Boniort *et al* (1980)
TeO_2–WO_3–Ta_2O_5	–	Crucible	–	Boniort *et al* (1980)
TeO_2–WO_3–Bi_2O_3 TeO_2–BaO–PbO				

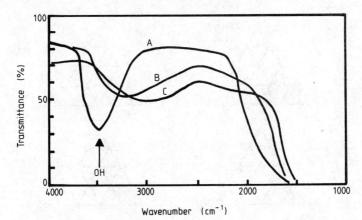

Figure 4.1 Infrared absorption spectra for GeO_2-based glasses (A, $92GeO_2$–$8Sb_2O_3$; B, $50GeO_2$–$50PbO$; C, $87.5TeO_2$–$12.5PbO$) (after Osawa *et al* 1980).

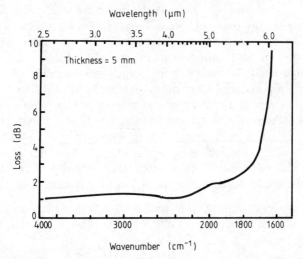

Figure 4.2 Infrared absorption spectra for $(TeO_2)_{70}ZnO_{20}BaO_{10}$ glasses (after Boniort *et al* 1980).

The glass forming regions of TeO_2-based glasses are shown in figure 4.3 (Boniort *et al* 1980). Refractive indices of GeO_2-based and TeO_2-based glasses are around 2. The wavelength dependence of the refractive index, which relates closely to material dispersion, is discussed in section 4.1.4.

4.1.3 Fabrication methods
GeO_2-based glass fibers have been fabricated by a VAD (vapor-phase

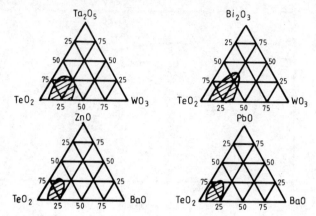

Figure 4.3 Glass forming regions for TeO₂-based glasses (after Boniort *et al* 1980).

axial deposition) process (Takahashi *et al* 1982, Takahashi and Sugimoto 1984). The VAD process is one of the most reliable processes for the fabrication of highly pure SiO₂-based glass fibers because of its excellent controlability and growth rate (Izawa 1977). A schematic view of the VAD process for the fabrication of the SiO₂-based glass fibers is shown in figure 4.4. In this method, germanium tetrachloride (GeCl₄) is used as the raw material. Antimony pentachloride (SbCl₅) is also used as the dopant raw material when the refractive index must be changed in the radial direction. GeCl₄ and SbCl₅ are bubbled with Ar gas and transferred into the oxyhydrogen flame where they are hydrolyzed into oxide particulates, which are deposited on the end of a starting substrate. The substrate is spun and lifted vertically at constant speed following the oxide deposit growth. A porous preform is sintered to the

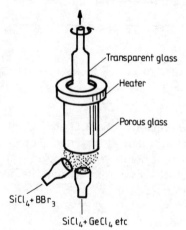

Figure 4.4 A schematic view of the VAD process for the fabrication of SiO₂-based glass fibers.

transparent preform in an electric furnace which is charged with helium gas. The helium gas plays an important role in eliminating the pores in the preform. The size of the sintered preform is normally 15 mm in diameter and 100 mm in length. The preform thus obtained is drawn using the fiber drawing machine with the electric furnace. The furnace temperature is held at about 1000 °C. A fiber is drawn to a length of at least 400 m and 150 μm diameter, and is then coated with silicon.

A conventional crucible technique can be used in the preparation of various heavy-metal oxide glass fibers. TeO_2-based glass fibers have been prepared with a gold crucible (Boniort *et al* 1980).

4.1.4 Properties
4.1.4.1 Transmission loss
The loss spectra for the GeO_2-based glass fibers were extensively studied by Takahashi and colleagues (Takahashi *et al* 1982, Takahashi and Sugimoto 1984). Figure 4.5 shows the loss spectrum for the fiber with a GeO_2 glass cladding. The fiber length is 400 m, and the outer diameter and core diameter are 150 μm and 70 μm, respectively. The numerical aperture *NA* is 0.11. In the figure, the absorption peaks at 1.4 and 2.2 μm are due to water. The minimum losses of 4 dB km^{-1} at 2 μm and 15 dB km^{-1} at 2.4 μm are obtained.

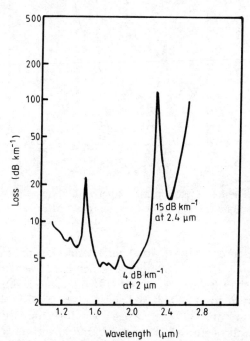

Figure 4.5 The loss spectrum for a fiber with a GeO_2–Sb_2O_3 glass core and GeO_2 glass cladding (after Sugimoto *et al* 1986).

Figure 4.6 shows, on the other hand, the loss spectrum for the fiber with a GeO_2 glass core and GeO_2–SiO_2 glass cladding (Sugimoto *et al* 1986). The fiber dimensions are 150 μm outer diameter and 100 μm core diameter. The numerical aperture *NA* is 0.17. The minimum loss of this fiber is almost the same as that of the Sb_2O_3-doped core fibers, although the base line of the transmission loss without impurity loss peaks is different for the fibers with SiO_2-doped cladding and with an Sb_2O_3-doped core. Transmission loss increases toward the shorter wavelength for the Sb_2O_3-doped core fiber; this is attributed to the electronic absorption of the Sb element.

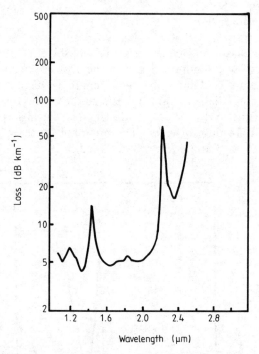

Figure 4.6 The loss spectrum for a fiber with a GeO_2 glass core and GeO_2–SiO_2 glass cladding (after Sugimoto *et al* 1986).

Analytical research indicates that an attenuation in transmission loss of less than 0.1 dB km^{-1} could be achieved around 2.2–2.4 μm wavelength (Olshansky and Scherer 1979). Figure 4.7 shows the theoretical loss spectrum for GeO_2-based glass fiber (Takahashi *et al* 1982). The loss spectrum for SiO_2 is also shown for comparison. However, another estimation of ultraviolet and infrared absorptions as well as the Rayleigh scattering revealed that the predicted minimum loss in GeO_2-

based glass fiber is close to 0.25 dB km^{-1} at $2 \mu m$ (Dianov 1982). In the latter estimation, the transmission loss of GeO_2-based glass optical fiber is not superior to that of conventional SiO_2-based glass fiber. Further research is however still required to clarify the minimum loss attainable in GeO_2-based glass fiber.

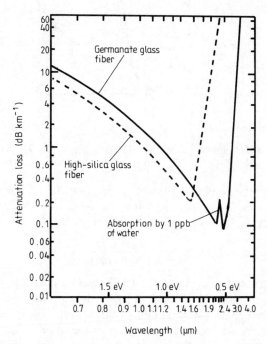

Figure 4.7 The theoretical attenuation loss spectrum of GeO_2 glass optical fiber, together with that already achieved with high-silica fiber (after Takahashi *et al* 1982).

The transmission loss of TeO_2-based glass fiber was measured by Boniort *et al* (1980). The loss spectrum for $200 \mu m$ diameter TeO_2-based glass fiber is shown in figure 4.8. This fiber is made from dehydrated glass by using a conventional drawing apparatus at 400 °C. It shows relatively high losses of 1 dB m^{-1} at $2 \mu m$ and 20 dB m^{-1} at $4 \mu m$.

4.1.4.2 Refractive index and dispersion
The refractive index behavior of the various glasses was discussed together with the other materials by Wemple (1977). Table 3.4 lists the refractive indices, which are found to be slightly higher than that of SiO_2 glass.

The material dispersions were extensively discussed by Nassau (1981).

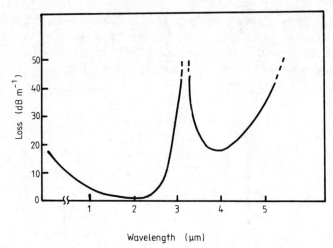

Wavelength (μm)

Figure 4.8 The loss spectrum for $(TeO_2)_{70}ZnO_{20}BaO_{10}$ glass optical fiber. The outer diameter is 200 μm (after Boniort *et al* 1980).

He calculated the wavelength where the material dispersion becomes zero. Based on the optical oscillator strengths and excitation energies, the wavelength λ_0 for which $d^2n/d\lambda^2 = 0$ is given as

$$\lambda_0 = hc \left(\frac{10^{-10}cn'f\mu d^3}{4\pi e^2 ZE^3} \right)^{1/4}, \qquad (4.1)$$

where E is the average electronic (Sellmeier) excitation gap (usually a few eV higher than the band gap), f is the normalized oscillator strength, Z is the formal valence of the anion, n' is the number of valence electrons on the anion in the compound, d is the anion–cation distance, μ is the reduced mass of the anion–cation pair, h is Planck's constant, c is the velocity of light, and e is the charge on the electron. At $n' = 8$ for closed-shell anions, equation (4.1) can be rewritten in practical units:

$$\lambda_0 = 2.96 \left(\frac{d^2 f\mu}{E^3 Z} \right)^{1/4}, \qquad (4.2)$$

where λ_0 is in μm, d is in Å, and E and f are in eV. Using equation (4.2), λ_0 can be calculated as shown in table 4.2. In the calculation, values of the required parameters were taken from Wemple's data (Wemple 1977). It can be understood from the table that λ_0 extends only up to about 5 μm.

Wood *et al* (1982) investigated optical properties of some oxide glasses. The glass compositions examined are GeO_2-based glasses: GeO_2 containing at least one of $BiO_{1.5}$, $TlO_{0.5}$, PbO, and $SbO_{1.5}$. These glasses are thought to be stable because of their high glass forming ability. Table 4.3 shows the compositions and some properties of the

Table 4.2 Optical parameters for various oxide glasses (after Nassau 1981).

Oxide	E	f	λ_0	Oxide	E	f	λ_0
MoO_3	6	4	2.8	Sc_2O_3	11	3	1.8
WO_3	6	4	3.0	Y_2O_3	9	4	2.4
P_2O_5	13.5	5	1.3	La_2O_3	8	4	2.9
As_2O_5	11	5	1.8	ZnO	6.1	3.1	2.7
Nb_2O_5	7	4	2.6	PbO	4.7	3.8	3.9
Ta_2O_5	7	4	2.7	BeO	13.7	5.1	1.2
SiO_2	13.3	5	1.3	MgO	11.4	2.8	1.6
GeO_2	11.0	4.0	1.7	CaO	9.9	2.8	2.0
SnO_2	8.1	3.0	2.2	SrO	8.3	2.5	2.5
TeO_2	6.3	3.9	2.8	BaO	7.1	2.4	3.0
TiO_2	5.5	3.8	2.8	Tl_2O	4	4	5.3
ZrO_2	11	4	2.0	Li_2O	12	5	1.6
HfO_2	10	4	2.2	Na_2O	11	4	2.1
ThO_2	10	4	2.3	K_2O	11	4	2.3
B_2O_3	12.4	6.8	1.3	Rb_2O	10	3	2.5
Al_2O_3	13.4	3.8	1.4	Cs_2O	10	3	2.7
Ga_2O_3	9.5	4	2.0				
In_2O_3	7	4	2.8	GaAs	3.7	4.5	6.3
As_2O_3	11	5	1.9	ZnTe	4.4	5.2	6.6
Sb_2O_3	7	5	2.9	BN	10.6	4.5	1.2
Bi_2O_3	5	5	3.9				

E = average electronic excitation gap, f = normalized oscillator strength, λ_0 = wavelength giving zero material dispersion.

Table 4.3 Compositions and some properties of GeO_2-based glasses (after Wood *et al* 1982).

Glass	A	B	C	D	E
Composition (mol.%)					
GeO_2	30	35	41	40	50
$BiO_{1.5}$	30	12	–	–	20
$TlO_{0.5}$	40	41	39	–	–
PbO	–	12	20	20	30
$SbO_{1.5}$	–	–	–	40	–
Annealing temp. (°C)	260	200	200	325	400
Annealing time (h)	3	3	3	1	16
Density (g cm^{-3})	7.66	7.35	6.98	5.40	5.99
Thermal expansion (10^{-6} K^{-1})	15.0	14.2	15.9	13.0	10.6
Glass transition temp. (°C)	270	260	240	350	470
Crystallization temp. (°C)	370,430	380	400	–	–

glasses (Wood *et al* 1982). Table 4.4 shows the optical properties of the glasses including refractive index and λ_0 wavelength (Wood *et al* 1982). In the table, $dM/d\lambda$ means the slope of the material dispersion, where $M = -\lambda/c(d^2n/d\lambda^2)$.

Table 4.4 Optical properties of the GeO_2-based glasses. The compositions A–E correspond to those shown in table 4.3 (after Wood *et al* 1982).

Glass	A	B	C	D	E
n_D	2.285 53	2.199 01	2.113 22	1.995 61	1.939 48
n_{λ_0}	2.141 75	2.074 84	2.006 48	1.925 66	1.881 82
Dispersion $= n_F - n_C$	0.122 26	0.103 79	0.085 76	0.051 15	0.041 37
Abbé number	10.51	11.55	12.98	19.46	22.71
Observed λ_0 (μm)	2.81	2.73	2.65	2.22	2.08
Predicted λ_0 (μm)	3.3	3.1	3.0	2.4	2.5
Observed $dM/d\lambda$ (ps nm^{-1} km^{-1} μm^{-1})	54.8	56.5	56.5	74.2	85.0
Predicted $dM/d\lambda$ (ps nm^{-1} km^{-1} μm^{-1})	27	26	27	~35	35

From the above results, it can be concluded that the wavelength giving zero material dispersion falls mainly in the range 2–3 μm. Fortunately, this wavelength range corresponds to that where the transmission loss is expected to be minimum. This coincidence between material dispersion and transmission loss is very important for long distance optical fiber communications.

4.1.4.3 Non-linear optical effects

The stimulated Raman scattering in GeO_2 glass fiber is interesting because the cross section of Raman scattering of GeO_2 glass is about nine times higher than that of SiO_2 glass (Galeener *et al* 1978). Takahashi and Sugimoto (1984) have confirmed this phenomenon using a GeO_2 glass fiber pumped by an Nd–YAG laser. Figure 4.9 shows the spectrum for stimulated Raman scattering in GeO_2 glass fiber (Sugimoto *et al* 1986). It can be seen from the figure that the ninth Raman line is obtained. The Raman scattering light can be used for the tunable light source which allows the dispersion properties of the optical fiber to be measured.

4.1.5 Summary

The minimum loss of heavy-metal oxide glass fibers so far reported is 4 dB km^{-1} at a 2 μm wavelength (Takahashi and Sugimoto 1984) for a

GeO_2 glass fiber fabricated by a vapor phase axial deposition (VAD). Therefore, GeO_2 glass fibers are thought to be good candidates for low loss optical fibers. Further reduction of the minimum loss depends strongly on the elimination of OH impurity. The VAD method incorporating a dehydration process has offered a reduction of OH impurity levels to as low as 1 ppb for SiO_2 glass fiber. However, it may be more difficult to achieve the OH reduction for GeO_2 glasses, because the chemical and physical properties are much more complicated than those of SiO_2 glasses. Therefore extensive studies will be required to reduce the OH impurity to the same level as that of SiO_2 glasses.

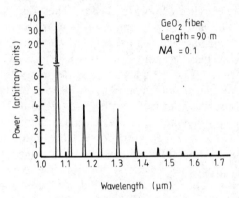

Figure 4.9 The spectrum for stimulated Raman scattering in GeO_2 glass fiber (after Sugimoto *et al* 1986).

Non-linear phenomena such as the stimulated Raman scattering using a GeO_2 glass fiber are very interesting for various applications. For example, the light created by stimulated Raman scattering can be used for the tunable light source which enables us to measure the dispersion properties of the optical fiber. Furthermore, GeO_2 glass fibers can be used in the transmission lines of infrared rays. One example of this application is in the transmission line of a radiometric thermometer. The wide infrared transmission band makes it possible to measure relatively low temperatures.

4.2 Fluoride glass fibers

4.2.1 Introduction
Fluoride glass fibers currently offer the best prospect for ultra-low loss optical fibers in the infrared region. Fluoride glasses have various desirable optical characteristics, such as a broad transparency range, low

refractive index, small dispersion, low Rayleigh scattering, and ultra-low thermal distortion. Furthermore, theoretical prediction shows that these fluoride glass fibers possess the lowest transmission loss of the infrared optical fibers composed of the glass materials, such as chalcogenide glasses and heavy-oxide glasses (Miyashita and Manabe 1982).

From the historical point of view, BeF_2-based glasses and then AlF_3-based glasses were synthesized in the early stages of research. However, these glasses have significant problems of toxicity (BeF_2-based glasses) and glass forming (AlF_3-based glasses), and so research on the fluoride glasses was limited.

However, in 1975, Lucas and Poulain made an important break-through in the research on fluoride glasses. They discovered a new glass system which exhibits a lesser tendency toward crystallization and a higher infrared transparency than the glasses so far fabricated (Poulain *et al* 1975). The glass system was based on heavy-metal fluoride, particularly ZrF_4. The typical glass compositions were 50–70 mol.% ZrF_4 as the primary constituent, 30 mol.% BaF_2 as a network modifier and a small amount of another metal fluoride, like ThF_4 or a rare-earth fluoride, playing the role of glass stabilizer. They have a relatively wide working range. Moreover, they have the advantage of high compositional flexibility, which allows them to be tailored to a broad range of properties essential for forming compatible core and cladding materials. These advantages led to increasing activity in the fabrication of the fluoride glass optical fibers.

4.2.2 Materials
4.2.2.1 Classification

Typical fluoride glasses are summarized in table 4.5. Of those shown only the BeF_2 glass can be vitrified without using a glass modifier. Furthermore, BeF_2 easily forms a glass on cooling from the molten state. On the other hand, ZrF_4-based and HfF4-based glasses exhibit a lesser tendency toward crystallization and a higher infrared transparen-cy. These ZrF_4 (or HfF_4) based glasses contain ZrF_4 (or HfF_4) as a glass network former (50–70 mol.%), BaF_2 as a primary network modifier (about 30 mol.%), and one or more additional metal fluorides of the rare-earths, alkalis, or actinides as glass stabilizers.

Fluorozirconate glasses, which are most resistant to devitrification, always include four or more fluoride components. They possess a relatively wide working range, which is defined here as the difference between the crystallization temperature T_x and the glass transition temperature T_g, of typically 100–150 °C, and thus can be cast into large preforms of high optical quality. The fluorozirconate glasses also show a usefully wide range of compositional flexibility.

Table 4.5 Typical fluoride glass compositions.

No.	Network former	Modifier	Glass stabilizer	Property
1.	BeF_2	–	–	Stable but hygroscopic and toxic
2.	AlF_3	BaF_2	CaF_2, YF_3	Tendency to devitrify
3.	ZrF_4	BaF_2	1. Rare-earth fluoride GdF_3, LaF_3	Stable
			2. Alkali fluoride LiF, NaF	Stable
			3. Actinide fluoride ThF_4	Stable
			4. Metal fluoride AlF_3, PbF_2, SbF_3, BiF_3	Stable
			5. Transition metal fluoride CdF_3	Stable
4.	HfF_4	BaF_2	LaF_3, AlF_3, PbF_2, CsF, ThF_4	Stable
5.	ZrF_4–HfF_4	BaF_2	–	Stable
6.	ThF_4–BaF_2	–	MnF_2, ZnF_2	Relatively unstable
7.	AlF_3–PbF_3	–	–	Relatively unstable
8.	Transition metal fluoride ZnF_2	BaF_2, ThF_4, YF_3, YbF_3, AlF_3		Relatively unstable

Glasses based on heavier HfF_4 are virtually chemically identical to ZrF_4. These glasses are more transparent at longer wavelength than the corresponding ZrF_4-based glasses. Thus, at a certain wavelength, the absorption coefficient for HfF_4-based glass should be lower than that for an analogous ZrF_4-based glass.

A number of other fluoride glass systems with ZrF_4 and HfF_4-free compositions have been reported, including AlF_3, ThF_4–BaF_2, AlF_3–PbF_3, and ZnF_2-based systems. These glasses, however, are less stable than ZrF_4- and HfF_4-based glasses.

The fluoride glass systems which were studied mainly for fabricating infrared fibers are listed in detail in table 4.6, together with the references.

4.2.2.2 Glass properties

ZrF_4-based and HfF_4-based glasses are currently the materials most used for optical fibers. Although binary systems such as the CaF_2–ZrF_4, SrF_2–ZrF_4, BaF_2–ZrF_4, and PbF_2–ZrF_4 systems yield fluorozirconate glasses, the glass forming ability is very poor (Mitachi *et al* 1983b).

Table 4.6 Fluoride glasses and fibers.

Class	Composition	Structure	Fabrication	Property	Reference
BeF_2	$BeF_2(40)-MgF_2(20)-SrF_2(20)-AlF_3(20)$	Unclad	–	–	Miyashita and Manabe (1982)
AlF_3	$AlF_3(43)-BaF_2(20)-CaF_2(20)-YF_3(17)$	Unclad	–	–	Kanamori (1982)
ZrF_4	$ZrF_4(50)-BaF_2(25)-ThF_4(25)$	–	–	–	Poulain et al (1977)
	$ZrF_4(57.5)-BaF_2(33.7)-ThF_4(8.8)$	–	–	–	Chen et al (1979)
	$ZrF_4(58-69)-BaF_2(28-38)-GdF_3(2-7)$	Unclad	Preform	0.48 dB m^{-1} $(3.39\ \mu m)$	Mitachi and Manabe (1980)
	$ZrF_4(60)-BaF_2(33)-ThF_4(7)$	–	RAP	–	Robinson et al (1980)
	Core: $ZrF_4(60.5)-BaF_2(31.7)-GdF_3(3.8)-AlF_3(4)$ Clad: $ZrF_4(59.2)-BaF_2(31)-GdF_3(3.8)-AlF_3(6)$	Clad	Casting	0.1 dB m^{-1} $(2.5\ \mu m)$	Mitachi et al (1981a)
	$ZrF_4(61.8)-BaF_2(32.3)-GdF_3(3.9)-AlF_3(2)$	Clad	Preform	0.37 dB m^{-1} $(2.55\ \mu m)$	Mitachi et al (1981c)
	$ZrF_4(60)-BaF_2(31)-GdF_3(4)-LaF_3(1)-AlF_3(2.5)-PbF_2(1.5)$	–	–	–	Poignant et al (1981)
	$ZrF_4(57.5)-BaF_2(33.5)-LaF_3(5.5)-AlF_3(3.5)$	Unclad	–	–	Ginther and Tran (1981)
	$ZrF_4(57)-BaF_2(36)-LaF_3(3)-AlF_3(4)$	Unclad	–	–	Drexhage et al (1981)
	Core: $ZrF_4(60)-BaF_2(19)-LaF_3(6)-NaF(15)$ Clad: $ZrF_4(57)-BaF_2(12)-LaF_3(6)-NaF(25)$	–	–	–	Ohsawa et al (1981)
	$ZrF_4(60)-BaF_2(33)-ThF_4(7)$	–	–	–	Harrington et al (1981)
	$ZrF_4(60)-BaF_2(35)-LaF_3(5)$	–	–	–	Bendow et al (1981a)
	$ZrF_4(58)-BaF_2(33)-ThF_4(9)$	–	–	–	Bendow et al (1981a)
	$ZrF_4(62)-BaF_2(33)-LaF_3(5)$	–	–	–	Bendow et al (1981b)
	$ZrF_4(57)-BaF_2(34)-LaF_3(5)-AlF_3(4)$	–	–	–	Lucas (1982)

Table 4.6 (*cont.*)

Class	Composition	Structure	Fabrication	Property	Reference
	Core: $ZrF_4(51)$–$BaF_2(16)$–$LaF_3(5)$–$AlF_3(3)$–$LiF(20)$–$PbF_2(5)$ Clad: $ZrF_4(53)$–$BaF_2(19)$–$LaF_3(5)$–$AlF_3(3)$–$LiF(20)$	Clad	Rotational casting	6 dB m^{-1} (3.6 μm)	Tran *et al* (1982a)
	$ZrF_4(51)$–$BaF_2(16)$–$LaF_3(5)$–$LiF(20)$–$AlF_3(3)$–$PbF_2(5)$	Unclad	Crucible	–	Tran *et al* (1982c)
	$ZrF_4(58)$–$BaF_2(15)$–$LaF_3(6)$–$AlF_3(4)$–$NaF(21)$	Unclad	–	–	Ohsawa *et al* (1982)
	$ZrF_4(62)$–$BaF_2(33)$–$ThF_4(5)$	–	–	–	Drexhage *et al* (1982a)
	$ZrF_4(58)$–$BaF_2(33)$–$ThF_4(9)$	–	–	–	Drexhage *et al* (1982b)
	Core: $ZrF_4(61.8)$–$BaF_2(32.3)$–$GdF_3(3.9)$–$AlF_3(2)$ Clad: $ZrF_4(59.2)$–$BaF_2(31.0)$–$GdF_3(38)$–$AlF_3(6)$	Clad	Built-in casting	21 dB km^{-1} (2.55 μm)	Mitachi and Miyashita (1982)
	$ZrF_4(60)$–$BaF_2(32)$–$GdF_3(4)$–$AlF_3(4)$ $ZrF_4(63)$–$BaF_2(27)$–$LaF_3(5)$–$AlF_3(5)$–LiF, BiF_3, PbF_2	–	–	–	Ohishi *et al* (1982) Tran *et al* (1982c)
	Core: $ZrF_4(58.1)$–$BaF_2(30.4)$–$GdF_3(3.7)$–$AlF_3(3.8)$–$SbF_3(4)$ Clad: $ZrF_4(56.9)$–$BaF_2(29.8)$–$GdF_3(3.6)$–$AlF_3(5.7)$–$SbF_3(4)$	Clad	Built-in casting	12 dB km^{-1} (2.55 μm)	Mitachi *et al* (1983a)
	Core: $ZrF_4(60.5)$–$BaF_2(31.7)$–$GdF_3(3.8)$–$AlF_3(4)$ Clad: $ZrF_4(58.6)$–$BaF_2(30.7)$–$GdF_3(3.7)$–$AlF_3(7)$	Clad	Built-in casting	8.5 dB km^{-1} (2.12 μm)	Mitachi *et al* (1983c)

Table 4.6 (*cont.*)

Class	Composition	Structure	Fabrication	Property	Reference
	ZrF_4–BaF_2–LaF_3–AlF_3–LiF–PbF_2	Teflon FEP clad	Preform	–	Tran et al (1983)
	Core: ZrF_4(53)–BaF_2(19)–LaF_3(5)–AlF_3(3)–LiF(20)–Cl, Br, I Clad: ZrF_4(53)–BaF_2(19)–LaF_3(5)–AlF_3(3)–LiF(20)	Clad (graded)	Reactive vapor transport	–	Tran et al (1984a)
	ZrF_4(53)–BaF_2(20)–LaF_3(4)–AlF_3(3)–NaF(20)	Unclad	Crucible	–	Mimura et al (1984)
	Core: ZrF_4(60.2)–BaF_2(31.2)–GdF_3(3.8)–AlF_3(3.8)–PbF_2(1) Clad: ZrF_4(59.4)–BaF_2(31.1)–GdF_3(3.8)–AlF_3(5.7)	Clad	Built-in casting	30 dB km^{-1} (2.8 μm)	Ohishi et al (1984a)
	Core: ZrF_4(60.2)–BaF_2(31.2)–GdF_3(3.8)–AlF_3(3.8)–PbF_2(1) Clad: ZrF_4(59.2)–BaF_2(31.1)–GdF_3(3.8)–AlF_3(5.7)	Clad (single mode)	Built-in casting	160 dB km^{-1} (3.28 μm)	Ohishi et al (1984b)
	ZrF_4–BaF_2–LaF_3–NaF–AlF_3	Clad	Rod-in-tube	67 dB km^{-1} (2.3 μm)	Ohsawa and Shibata (1984)
	ZrF_4(62)–BaF_2(33)–LaF_3(5)-based glass	–	–	–	Poulain and Saad (1984)
	Core: ZrF_4(56)–BaF_2(30)–LaF_3(5)–ThF_4(4)–AlF(5) Clad: ZrF_4(55)–BaF_2(31)–LaF_3(5)–NaF(4)–AlF_3(5)	Clad	Preform	16 dB km^{-1} (2.6 μm)	Maze et al (1984)
	ZrF_4(51.5)–BaF_2(19.5)–LaF_3(5.3)–LiF(18.0)–AlF_3(3.2)–PbF_2(2.5)	Clad	Preform	–	France et al (1984)

Table 4.6 (*cont.*)

Class	Composition	Structure	Fabrication	Property	Reference
	Core: ZrF$_4$–BaF$_2$–NaF–AlF$_3$ Clad: ZrF$_4$–BaF$_2$–NaF–LaF$_3$–AlF$_3$–HfF$_4$	Clad	Preform	–	Tokiwa *et al* (1985a,b)
	ZrF$_4$(53)–BaF$_2$(20)–NaF(20)–LaF$_3$(4)–AlF$_3$(3)	–	Preform	–	Tokiwa *et al* (1985c)
	Core: ZrF$_4$(60.48)–BaF$_2$(31.68)–GdF$_3$(3.84)–AlF$_3$(4) Clad: ZrF$_4$(59.22)–BaF$_2$(31.02)–GdF$_3$(3.76)–AlF$_3$(6)	Clad	Built-in casting	10 dB km^{-1} (2.7 μm)	Mitachi *et al* (1985)
	ZrF$_4$(57–53)–BaF$_2$(30–20)–NaF(20–6)–LaF$_3$(4)–AlF$_3$(3)	–	Casting	–	Nakai *et al* (1985a,b, 1986a)
	ZrF$_4$–BaF$_2$–LaF$_3$–AlF$_3$–NaF–PbF$_2$	Clad	Rotational casting	21 dB km^{-1} (2.7 μm)	France *et al* (1985)
	ZrF$_4$(57)–BaF$_2$(36)–LaF$_3$(3)–AlF$_3$(4)	–	–	–	Brown and Hutta (1985)
	ZrF$_4$(55.8)–BaF$_2$(14.4)–LaF$_3$(5.8)–AlF$_3$(3.8)–NaF(20.2)	–	–	–	Brown and Hutta (1985)
	Core: ZrF$_4$(54)–BaF$_2$(30)–LaF$_3$(4.5)–AlF$_3$(2.5) Clad: ZrF$_4$(53)–BaF$_2$(24)–LaF$_3$(4)–AlF$_3$(4)–NaF(15)	Clad (single mode)	Rod-in-tube	>100 dB km^{-1}	Monerie *et al* (1985)
	ZrF$_4$(51)–BaF$_2$(16)–LaF$_3$(5)–LiF(20)–AlF$_3$(3)–PbF$_2$(5)	Unclad	Preform	–	Lau *et al* (1985)
	ZrF$_4$(57)–BaF$_2$(34)–LaF$_3$(5)–AlF$_3$(4)	Clad	–	–	Fonteneau *et al* (1985)
	Core: ZrF$_4$(53)–BaF$_2$(20)–NaF(20)–LaF$_3$(4)–AlF$_3$(3) Clad: ZrF$_4$(39.7)–HfF$_4$(13.3)–BaF$_2$(18)–NaF(22)–LaF$_3$(4)–AlF$_3$(3)	–	Double crucible in NaF	–	Nakai *et al* (1986a)

Table 4.6 (*cont.*)

Class	Composition	Structure	Fabrication	Property	Reference
	Core: $ZrF_4(49)$–$BaF_2(25)$–$LaF_3(3.5)$–$YF_3(2)$–$AlF_3(2.5)$–$LiF(18)$ Clad: $ZrF_4(50.6)$–$BaF_2(24.7)$–$LaF_3(2.6)$–$YF_3(2.6)$–$AlF_3(3.7)$–$LiF(6)$–$NaF(9.8)$	Clad (single mode)	Built-in casting	$8.5\,dB\,km^{-1}$ $(2.2\,\mu m)$	Ohishi et al (1986)
	ZrF_4–BaF_2–LaF_3–AlF_3–NaF	Clad	–	$4\,dB\,km^{-1}$ $(2.5\,\mu m)$	Lu et al (1986)
	Core: $ZrF_4(49)$–$BaF_2(25)$–$LaF_3(3.5)$–$YF_3(2)$–$AlF_3(2.5)$–$NaF(18)$ Clad: $ZrF_4(23.7)$–$HfF_4(23.8)$–$BaF_2(23.5)$–$LaF_3(2.5)$–$YF_3(2)$–$AlF_3(4.5)$–$NaF(20)$	Clad	Built-in casting	$0.7\,dB\,km^{-1}$ $(2.63\,\mu m)$	Kanamori and Sakaguchi (1986)
HfF_4	$HfF_4(62)$–$BaF_2(15)$–$LaF_3(5)$–$AlF_3(2)$–$PbF_2(10)$–$CsF(6)$	Plastic clad	–	–	Drexhage et al (1981)
	$HfF_4(62)$–$BaF_2(33)$–$LaF_3(5)$	–	–	–	Bendow et al (1981b)
	$HfF_4(57.5)$–$BaF_2(33.7)$–$ThF_4(8.8)$	–	–	–	Drexhage et al (1982a)
	$HfF_4(62)$–$BaF_2(33)$–$ThF_4(9)$	–	–	–	Drexhage et al (1982a)
ZrF_4–HfF_4	Core: $ZrF_4(27)$–$HfF_4(27)$–$BaF_2(23)$–$ThF_4(8)$–$LaF_3(4)$–$AlF_3(2)$–$LiF(3)$–$NaF(3)$–$PbF_2(3)$ Clad: $ZrF_4(26)$–$HfF_4(26)$–$BaF_2(23)$–$ThF_4(8)$–$LaF_3(4)$–$AlF_3(4)$–$LiF(4)$–$NaF(4)$–$PbF_2(1)$	Clad	Preform	$2\,dB\,m^{-1}$ $(2.5,\ 3.6\,\mu m)$	Poignant et al (1982)
ThF_4–BaF_2	ThF_4–BaF_2–MnF_2, ZnF_2	–	–	–	Fonteneau et al (1980)
AlF_3–PbF_2	AlF_3–PbF_2	–	Roller quenching	–	Kanamori et al (1980)
	PbF_2–$AlF_3(30–60)$	–	Rapid quenching	–	Shibata et al (1980a)

Table 4.6 (*cont.*)

Class	Composition	Structure	Fabrication	Property	Reference
ZnF_2	$ZnF_2(29)$–$BaF_2(20)$–$ThF_4(22.2)$–$YF_3(14.4)$–$AlF_3(14.4)$	–	–	–	Lucas (1982)
	$ZnF_2(26.5)$–$BaF_2(17.5)$–$YbF_3(26)$–$ThF_4(30)$	–	–	–	Drexhage et al (1982b)
	$ZnF_2(28)$–$BaF_2(15)$–$ThF_4(29)$–$YbF_3(28)$	–	–	–	Fonteneau et al (1985)
	$ZnF_2(28.3)$–$BaF_2(15)$–$ThF_4(28.4)$–$YbF_3(28.3)$	–	RAP	–	Tregoat et al (1986)

However, glass formation in the fluorozirconate systems can be remarkably improved by adding appropriate metal fluorides to the binary system glasses: a suitable combination of glass modifying fluorides brings about the stabilization of fluorozirconate glass. If fluorozirconate glasses are assumed to be ionic glass, then such a stabilization may be interpreted as the effective fixation of a fluoride ion random array by metal ions. In order to stabilize the random array of fluoride ions by metal ions, several metal ions of very different ionic radii are required and, at the same time, metal ions which admit various coordination environments (coordination number and configuration) are desirable. Ions such as heavy-metal elements and rare-earth elements would thus be preferable for the stabilization of fluorozirconate glass.

ZrF_4–BaF_2–GdF_3 systems are one of the most preferable choices for low loss optical fiber (Mitachi *et al* 1983b). Figure 4.10 shows the glass forming and fiber drawing regions in the ternary ZrF_4–BaF_2–GdF_3 system. The system containing GdF_3 has a larger glass forming area than other fluoride containing systems. The fiber drawing region is defined by drawing homogeneous fiber without devitrification by a conventional preform method. The glass transition temperature T_g and the crystallization temperature T_x are 310 °C and 376 °C, respectively. The deformation temperature T_d is 332 °C. The thermal expansion coefficient varies from 146 to 206 ($\times 10^{-7}/$°C). Since the fluorozirconate glasses contain only cations of large mass number, their density is quite high, $4.6 \, \mathrm{g\,cm^{-3}}$.

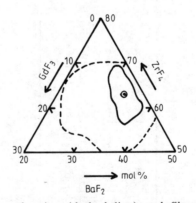

Figure 4.10 Glass forming (dashed line) and fiber drawing (solid line) regions in the ZrF_4–BaF_2–GdF_3 system. The region marked ⊙ is the composition used for fiber drawing (after Miyashita and Manabe 1982, © 1982 IEEE).

Table 4.7 summarizes the physical properties for ZrF_4–BaF_2–GdF_3 glass. A typical transmission spectrum for a bulk sample is shown in figure 4.11. The optical transparency interval for 50% transmittance is about 0.23–7 μm with no other absorption bands between these limits.

Table 4.7 Optical and thermomechanical properties of $ZrF_4(63)$–$BaF_2(33)$–$GdF_3(4)$ glass (after Miyashita and Manabe 1982, © 1982 IEEE).

Transmission range (μm)	0.23–7.0 (4 mm thick)
Refractive index (n_D)	1.529
Thermal expansion coefficient (°C^{-1})	175×10^{-7}
Glass transition temperature (°C)	310
Deformation temperature (°C)	332
Crystallization temperature (°C)	376
Density (g cm^{-3})	4.6

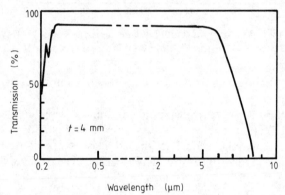

Figure 4.11 Optical transmission in ZrF_4–BaF_2–GdF_3 glass (after Miyashita and Manabe 1982, © 1982 IEEE).

The other important fluorozirconate glass is ZrF_4–BaF_2–LaF_3. Figure 4.12 shows the glass forming regions for ZrF_4–BaF_2–LaF_3 ternary glass. Note that only a small part in the center of this glassy area corresponds to a stable glass. However, most zirconium fluoride glasses can be stabilized by adding a small amount of aluminum fluoride. One example of the stable glass composition is therefore $ZrF_4 : BaF_2 : LaF_3 : AlF_3 = 57 : 34 : 5 : 4$ (Lucas 1982).

The ZrF_4–BaF_2–ThF_4 glass system is also important because it is relatively stable. The glass forming region for the ZrF_4–BaF_2–ThF_4 ternary system is shown in figure 4.13 (Poulain *et al* 1977). The optical transmission range is from 0.22 to 7 μm. The refractive index is about 1.53 and can be modulated by changing the chemical composition. In the system $ZrF_4 : BaF_2 : ThF_4 = 58 : 33 : 9$, the glass transition temperature and crystallization temperature are 320 °C and 420 °C, respectively.

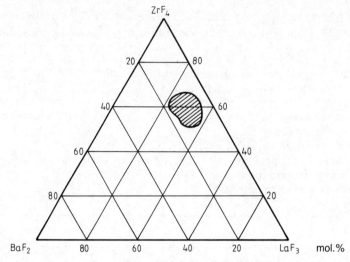

Figure 4.12 Glass forming regions for ZrF_4–BaF_2–LaF_3 ternary glass (after the catalogue of the Company Le Verre Fluore (LVF)).

Other ZrF_4-based glass systems have been examined, as shown in table 4.6.

HfF_4-based glasses are transparent to longer wavelengths than the corresponding ZrF_4-based glasses. As shown in figure 4.14, the multi-phonon absorption edges for HfF_4-based glasses are shifted to longer wavelengths with respect to the ZrF_4-based glasses (Drexhage *et al* 1982a).

4.2.2.3 Glass structure

The internal structure of ZrF_4–based glass can be described simply as follows (Mitachi *et al* 1983b). It is well known that ZrF_4 itself cannot be formed in the glass state, but can be formed by adding glass modifiers such as BaF_2.

In that case, the glass composition is typically $35 \leq BaF_2 \leq 40$, $60 \leq ZrF_4 \leq 65$ (mol.%). Figure 4.15 shows the proposed microstructure of the ZrF_4–BaF_2 glass. Since the coordination number of ZrF_4 is eight, the bonding of Zr–F forms a periodic structure as shown in figure 4.15(*a*). Once BaF_2 is introduced in the ZrF_4, the periodic structure is destroyed, and then the random structure is formed as shown in figure 4.15(*b*). This structure leads to the glass state of ZrF_4–BaF_2, which lacks long range periodicity. Furthermore, when the mole ratio of BaF_2 to ZrF_4 approaches unity, another periodicity in the structure appears, as shown in figure 4.15(*c*), resulting in an increase in the tendency toward crystallization. In addition, when GaF_3 is further

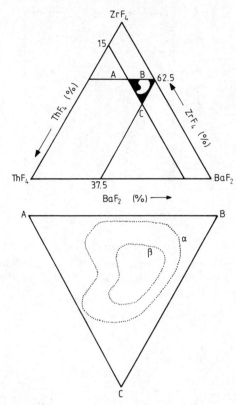

Figure 4.13 Glass forming regions for ZrF_4–BaF_2–ThF_3 ternary glass. α and β represent the regions obtained by glass cooling on the surface of metal maintained at 20 °C (α) and 300 °C (β), respectively (after Poulain *et al* 1977).

added to the structure in figure 4.15(*b*), randomness of the internal structure increases because the coordination number of GdF_3 is six. Therefore, the addition of GdF_3 stabilizes the glass structure of ZrF_4-based glass. The glass forming mechanism of other heavy-metal fluoride glasses such as HfF_4-based glass may be similar to that of the ZrF_4-based glass.

This glass forming mechanism suggests that in order to obtain a stable glass structure, it is necessary to select appropriate glass stabilizers which destroy the internal structure.

4.2.3 Fabrication methods
4.2.3.1 Bulk glass fabrication
In the fluoride glasses, BeF_2 easily forms a glass state upon cooling from

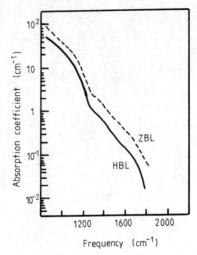

Figure 4.14 Multiphonon absorption coefficients for fluoride glasses: ZBL, ZrF_4–BaF_2–LaF_3 system; HBL, HfF_4–BaF_2–LaF_3 system (after Drexhage *et al* 1982a).

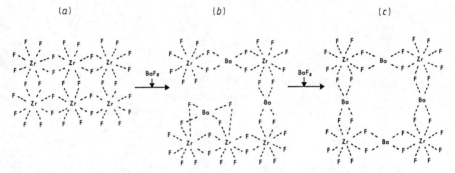

Figure 4.15 The proposed microstructure of ZrF_4–BaF_2 glass (Mitachi *et al* 1983b).

the melt. However, much attention is required for fabricating multi-component fluoride glasses such as ZrF_4–BaF_2–GdF_3, because they tend to have low viscosities at their liquidus temperature and a tendency toward crystallization. Moreover, fluoride melts are reactive with the atmosphere and with certain crucible materials. Therefore nonreactive crucible materials, such as vitreous carbon, platinum or gold are required. Stringent atmospheric control during melting and quenching is necessary to prevent oxide and hydroxide contamination which leads to undesirable absorption and possibly nucleation and crystallization (France *et al* 1984, Ohishi *et al* 1984a). Either inert atmospheres (N_2, Ar, He) or reactive gases such as CCl_4, SF_6, HF, CF_4, and BF_3 have been used to remove water and various oxide impurities from the melt. In many instances both the batching and weighing of starting materials, and the melting and casting, have been carried out inside glove boxes or in other controlled atmosphere chambers.

As shown in the review paper by Tran *et al* (1984b), glass melts may be obtained directly by fusing anhydrous fluorides at temperatures ranging from 800 to 1000 °C for time periods of the order of an hour, or by converting oxide raw materials to fluorides using ammonium bifluoride ($NH_4F \cdot HF$) at around 400 °C prior to the fusion process (Lucas *et al* 1978). The $NH_4F \cdot HF$ fluorination of oxides, for example ZrO_2, takes place as follows:

$$2ZrO_2(s) + 7NH_4F \cdot HF(s) \rightarrow 2(NH_4)_3ZrF_7(s) + NH_3(g) + 4H_2O(g)$$

$$(NH_4)_3ZrF_7(s) \rightleftarrows ZrF_4(s) + 3NH_4F.$$

Certain glass compositions may be cast by pouring the melts into room temperature or heated molds, or they may be formed directly in their crucible or in a sealed tube used for melting.

ZrF_4-based glasses have also been fabricated using a method in which the starting materials are molded in a die under a high pressure (Robinson *et al* 1980). In this method, molding is accomplished with a 20 ton hydraulic press. A 0.5 in. cylindrical tungsten carbide die is used. The die with sample inside is placed in the press, and the temperature is raised to 312 °C while increasing pressure is applied until a pressure drop due to glass flow is indicated. The temperature is held constant until the die is filled, as indicated by no further pressure drop even as higher pressure is applied. The temperature is then lowered to below 300 °C, the pressure is released, and the sample is cooled to room temperature. The glass fabricated by this method is continuously transparent from 0.3 to 7 μm.

Alternatively, if fabrication is difficult, the glass can be made using rapid quenching. PbF_2–AlF_3 fluoride glasses have been vitrified in the shape of a ribbon using roller quenching over the composition range 30–60 mol.% AlF_3 (Shibata *et al* 1980a, b). In this method, the material is melted in a platinum crucible which has an orifice of 0.4 mm in diameter. At 1100 °C the melt is forced out through the orifice with 0.8 kg cm^{-3} Ar gas pressure, onto the surface of a roller rotating at 2000 rpm. The molten jet is solidified extremely rapidly when it touches the roller surface—the quenching rate is about 5×10^5 °C s^{-1}. The sample thickness is 10–20 μm.

4.2.3.2 Purification

Purification is very important in fabricating low loss fluoride glass fibers. In particular, purification of the starting materials plays an important role in reducing the level of impurities in the glass.

Mitachi *et al* (1984) studied various impurity reduction methods in fabricating ZrF_4–BaF_2–GdF_3–AlF_3 glass fibers. Purification methods using recrystallization, extraction, chemical vapor deposition, distillation and sublimation were studied systematically.

In the recrystallization method for ZrF_4, zirconium carbonate $Zr(CO_3)_2$ is used as a starting material, which becomes zirconium hydroxide $Zr(OH)_4$ and then zirconium ammonium fluoride $(NH_4)_3ZrF_7$. The process can be described as:

$$Zr(CO_3)_2 \rightarrow ZrOCl_2 \rightarrow Zr(OH)_4 \rightarrow (NH_4)_3ZrF_7 \rightarrow ZrF_4.$$

The recrystallization is carried out in the ZrF_4 synthesis process, that is $ZrOCl_2$ and $(NH_4)_3ZrF_7$ are each recrystallized in aqueous solution three times. Finally, recrystallized $(NH_4)_3ZrF_7$ is decomposed into ZrF_4 and NH_4F by heating at 450 °C in a platinum crucible, whereupon NH_4F is driven off and purified ZrF_4 is obtained. This method can reduce impurity levels in ZrF_4 to below 1 ppm.

Since fluorides of transition metal impurities are much more soluble in hydrochloric acid solution than ZrF_4, BaF_2, GdF_3 and AlF_3, the starting materials can be purified by an extraction process using HCl. For example, ZrF_4, BaF_2, GdF_3, and AlF_3 are individually settled in the extractor and transition metal impurities are repeatedly extracted out with the distilled 12% HCl solution for 4 h.

In order to use chemical vapor deposition, starting materials that can be vaporized are required. However, there are only a few reagents, as a source of Zr^{4+}, Ba^{2+}, and Gd^{3+}, which easily vaporize. Zirconium tertiarybuthoxide $Zr(OC_4H_9)_4$ and barium neodecanoate $Ba(C_{10}H_{19}O_2)_2$ can however be used as starting materials. They react with HF gas, and the reaction seems to proceed as follows:

$$Zr(OC_4H_9)_4 + 4HF \rightarrow ZrF_4 + 4C_4H_9OH$$

$$Ba(C_{10}H_{19}O_2)_2 + 2HF \rightarrow BaF_2 + 2C_{10}H_{20}O_2.$$

The fine particles of ZrF_4 and BaF_2 thus produced are collected. However, the yield is extremely low, since the vapor pressure of $Zr(OC_4H_9)_4$ is only 30 mmHg at 127 °C and it is not stable at boiling temperature. Mitachi *et al* (1984) therefore concluded that the chemical vapor deposition method is inappropriate for zirconium fluoride glass fiber purification.

In the distillation method, BaF_2 and GdF_3 were melted in platinum boats which were settled in silica tubes at 1280 and 1180 °C, respectively, under 3 mmHg dry Ar atmosphere. However, such a high temperature deforms the silica vessel, resulting in some difficulty in distillation. Therefore the distillation method can be used only if a suitable vessel for high temperature use is found, which would guarantee sufficient purity and resistivity against fluoride attack.

Figure 4.16 shows a diagram of the sublimation system. Raw materials such as ZrF_4, AlF_3 and $NH_4F \cdot HF$ are settled in platinum boats. They are sublimated at 900 °C in a dry Ar atmosphere of 1–3 mmHg and collected on a cylindrical plate. The typical sublimation rate is 0.3 g min^{-1}. AlF_3 is also sublimated at 1000 °C under reduced pressure. The fluorinating material $NH_4F \cdot HF$ is sublimated at 200 °C in a dry Ar atmosphere of 3 mmHg. On the other hand, BaF_2 and GdF_3 are purified by sublimating out transition metal impurities at 1180 °C in a dry Ar atmosphere of 3 mmHg. This sublimation method is found to be most effective for preparing fluoride materials with impurity levels of less than 0.1 ppm.

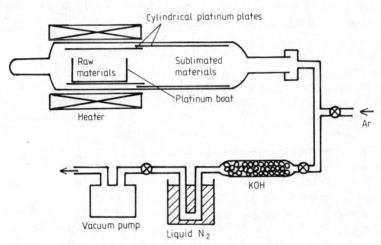

Figure 4.16 The sublimation process (after Mitachi *et al* 1984; © 1984 IEEE).

Robinson *et al* (1980) have reported a reactive atmosphere processing (RAP) technique. In this technique, starting materials are purified using vitreous carbon or platinum crucibles in a furnace. The furnace is vacuum pumped to 300 °C prior to RAP. The reactive atmosphere is either anhydrous HF diluted to 10 mol.% with He, or gaseous CCl_4 diluted with He. The mixture of starting materials is heated to 900 °C in 5 h; the resulting melt is soaked for 5 h and then cooled abruptly to room temperature. The ingot thus obtained can then be annealed at 220 °C to remove strain.

A reactive atmosphere processing using NF_3 gas has been reported by Nakai *et al* (1986b). The hydroxyls and other complex ions such as NH_4^+ or SO_4^{2-} can be effectively eliminated by conducting RAP with NF_3, without any resultant increase in scattering. Using this method, an absorption loss due to iron ions can also be reduced (Nakai *et al* 1985b). This is because of the conversion of ferrous to ferric ions by RAP with NF_3. In fluoride glasses, both ferric and ferrous ions can exist, but a large proportion of the ions are ferrous. They have absorption peaks at 1.1 μm and 1.8 μm and cause a serious loss increase in the minimum loss region (2–4 μm). On the other hand, ferric ions, whose absorption peaks lie around 0.5 μm, have less influence on the 2–4 μm range. Thus the oxidation of ferrous to ferric ions reduces the absorption loss in the minimum loss region, even if the iron in the glass is not actually eliminated. The reduction of the absorption loss due to iron impurities by oxidation has also been reported by France *et al* (1985).

Maze *et al* (1984) have reported the distillation method using only the thermal procedure, which allows them to obtain a substantial reduction of OH absorption in fluoride glasses.

4.2.3.3 Glass fiber fabrication

Glass fiber fabrication methods can be basically classified into two distinct categories: one is the rod drawing method that is sometimes called the preform method, and the other is the crucible method that enables us to draw glass fibers directly through a nozzle on the crucible.

The preform method. The simplest preform method is the one which produces glass fibers with uniform refractive indices along the axial direction. These fibers are called the unclad fibers—the air acts as the cladding material which reflects light rays at the boundary.

In general, fluoride glasses have a low deformation temperature compared to the conventional silica glasses, so that it is possible to draw fibers at temperatures below 400 °C. However, viscosities of fluoride glasses depend strongly on temperature, so that precise temperature control is needed to obtain high quality fibers.

The glass rod may be jacketed with an appropriate plastic such as Teflon FEP (Mitachi *et al* 1981c). The glass rod with Teflon jacket can be heated in a furnace and drawn directly into a continuous length of plastic-clad fiber. The apparatus for drawing is shown in figure 4.17. This method has several advantages. First, the simultaneous drawing of the jacket enables us to obtain a fiber whose surface has few micro-cracks, and which is therefore stronger. Secondly the plastic cladding makes it possible to transmit light rays stable because the jacket protects the light from the surrounding undesirable environment.

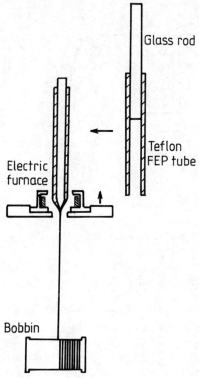

Figure 4.17 The apparatus for drawing an optical fiber jacketed with Teflon FEP (after Mitachi *et al* 1981c).

Optical fibers with cladding can also be fabricated by the following methods:

(i) built-in casting
(ii) rotational casting
(iii) reactive vapor transport.

Mitachi *et al* (1981b) developed the built-in casting method. Figure 4.18 shows the built-in casting process schematically. A cladding glass melt is poured into a cylindrical brass mold preheated to around the glass transition temperature. This mold is turned over so that the glass melt in the central part of the mold runs out, and then a core glass melt is poured into it and annealed. The typical preform obtained has a 10 mm outer diameter, and is 120 mm long. This method is advantageous because the boundary between the core and the cladding is very smooth, and therefore scattering loss can be reduced.

Tran *et al* (1982a) modified the casting technique to achieve improved

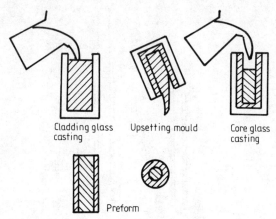

Figure 4.18 Schematic of a built-in casting process for a fluoride glass preform (after Mitachi *et al* 1982).

uniformity and interface quality. The schematic representation is shown in figure 4.19. In this method highly concentric glass tubes of precise thickness can be made by pouring a glass melt into a gold-coated cylindrical mold preheated to glass transition temperature, which is then spun at speeds of around 3000 rpm or more. A core glass melt is then poured into it to obtain a glass preform with a well defined interface.

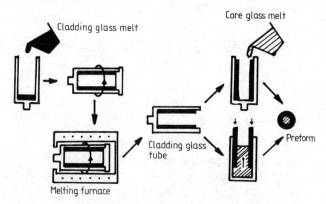

Figure 4.19 A schematic representation of the steps required for the rotational casting of fluoride glass preforms (after Tran *et al* 1982a).

Tran *et al* (1984a) have also reported a vapor phase technique called reactive vapor transport (RVT) for the fabrication of fluoride glass preforms. A reactive mixture of vapors, originating from low vapor pressure metal halides or halogenated gases, is transported through a rotating fluoride glass tube. Proper control of the process parameters results in substantial ion exchange and incorporation of Cl, Br or I ions into the glass, thereby raising the refractive index and forming a core region. A cross section of a fluoride glass tube preform, illustrated in figure 4.20, shows a ZrF_4–BaF_2–LaF_3–AlF_3–LiF substrate tube with the core region doped with Cl, Br, or I via the RVT process. It has been demonstrated that a core–cladding index difference of at least 0.004 can be achieved, and that the core region can be varied from 0.2 to ≥ 3 mm. The refractive index profile was a parabolic variation due to a diffusion-controlled process. The RVT method thus offers potential for achieving graded index fibers, as well as reducing OH contamination since the vapors are reactive.

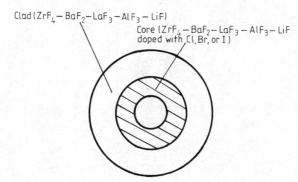

Figure 4.20 A cross-sectional representation of a fluoride glass tube preform prepared by a reactive transport process (after Tran *et al* 1984a).

Of course, the conventional rod-in-tube method can be used for the fabrication of the fluoride core-clad fiber. However, it must be noted that some attention is required if a smooth core-clad interface is to be obtained.

The crucible method. The advantage of the crucible approach is that it is a continuous process. Moreover, crystal formation can be substantially eliminated if the fluoride glass melt can be rapidly quenched into glass fibers, as was shown by Tran *et al* (1984b). However, Tran *et al* also suggested that since the low viscosity of these glasses (<10 P) still persists at slightly above the drawable temperature range, extreme precautions and stringent crucible design are required to avoid devitrification and fiber distortion during drawing.

Drexhage *et al* (1981) reported a single-crucible apparatus used to prepare fluoride glass fibers, as shown in figure 4.21. It consists of a high purity platinum crucible with a lid and a tapered downpipe held within a resistance-heated furnace. The crucible is charged with a premelted cullet. Separate thermal control is provided for the nozzle assembly. The design permits the use of inert or reactive (e.g. N_2 or CCl_4) atmosphere above the melt. A cooling fixture is employed to quench the glass as it leaves the downpipe orifice. The glass fiber passes through a coating applicator prior to being wound on a drum. Lengths of plastic clad fiber many metres long are obtained. HfF_4–BaF_2–LaF_3 and ZrF_4–BaF_2–LaF_3 based fluoride glass fibers have been prepared in this way.

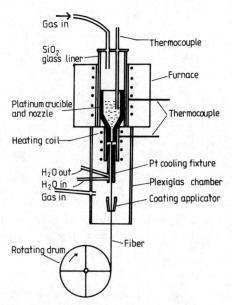

Figure 4.21 The single-crucible apparatus used to make fluoride glass fibers (after Drexhage *et al* 1981).

Tran *et al* (1982b) succeeded in preparing fibers 1 km long from a ZrF_4–BaF_2–LaF_3–AlF_3–LiF–PbF_2 glass system using a specially designed crucible. Fiber drawing is carried out using a two-crucible drawing set-up. Bulk fluoride glasses in a specially designed Pt crucible are preheated to a molten state and are transferred to a gradient-free furnace heated at the glass softening temperature for fiber drawing. Alternatively, as shown in figure 4.22, the molten fluoride glasses are drained from an upper melting crucible to a lower single drawing crucible maintained at the glass softening temperatures prior to fiber drawing. However, the glass fibers thus obtained showed significant optical losses because of the presence of various scattering defects, such as bubbles, fiber non-uniformities, and microcrystallites.

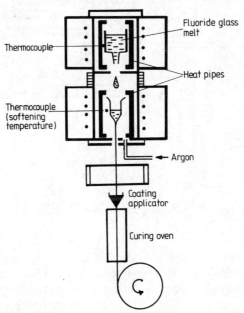

Figure 4.22 The double-stage crucible method for the fiber drawing of fluoride glasses (after Tran *et al* 1982b).

Mimura *et al* (1984) developed a novel crucible method. The proposed apparatus is shown in figure 4.23. In this system, glasses are melted in a gold crucible inserted in an airtight graphite vessel. After sufficient melting, the melt is transferred onto an aluminium disc with a small hole at its center through the gold pipe by controlling the gas pressure in the graphite vessel. The melt is cooled on the aluminium disc and then drawn into fibers. Unclad $ZrF_4(53)$–$BaF_2(20)$–$LaF_3(4)$–$AlF_3(3)$–$NaF(20)$ glass fibers of 130 μm diameter and 200 m length were drawn without crystallization. The drawing speed is 6 m min^{-1}. During the drawing, the temperature of the aluminium disc is held at the flowing point of the glass. The glasses are cooled at a rate exceeding 5 °C min^{-1} during the drawing process, so that fluoride glass fibers can be fabricated which are free from crystallization. The double-crucible technique based on the same idea by Mimura *et al* (1984) was also reported by Tokiwa *et al* (1985a). 500 m long single-mode and 700 m long multimode fibers were obtained using ZrF_4–BaF_2–NaF–LaF_3–AlF_3 core and ZrF_4–BaF_2–NaF–LaF_3–HfF_4 clad glasses.

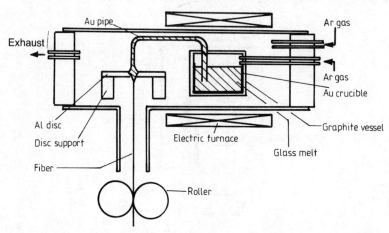

Figure 4.23 The fiber drawing system for the crucible method (after Mimura *et al* 1984).

Nakai *et al* (1986a) recently reported a method for the preparation of fluoride glass preforms under an NF_3 atmosphere. The technique uses a double crucible and all stages of the fabrication process are carried out under an NF_3 atmosphere, which can effectively eliminate impurity absorption.

In general, at present, low loss fluoride glass fibers are obtained mainly by the preform method, because of the excellent smoothness of the core-clad interfaces. Efforts have therefore been concentrated on this method.

4.2.4 Properties

4.2.4.1 Loss characteristics

Transmission losses of various optical fibers. The transmission loss of fluoride glass fiber is now rapidly decreasing as a result of extensive study of the loss reduction mechanism, and the loss value reported here will probably be reduced further in the near future. The minimum loss obtained to date is $0.7 \, dB \, km^{-1}$ (Kanamori and Sakaguchi 1986) and $0.9 \, dB \, km^{-1}$ (Tran 1986) at a wavelength of around $2.5 \, \mu m$. These values are the same order of magnitude as that of low loss silica glass fiber. In the following, the loss characteristics of various fluoride glass fibers are described in detail.

The transmission losses are different for unclad glass fiber, including

Teflon-jacketed fiber, and for core-clad glass fiber. In general, the transmission loss of unclad fiber is influenced by the surrounding materials, such as air or Teflon. On the other hand, the core-clad type glass fiber is barely influenced by external circumstances. Figure 4.24 shows a typical example of a transmission loss spectrum for a Teflon FEP clad glass fiber, which is drawn from a $ZrF_4(60.5)–BaF_2(31.7)–GdF_3(3.8)–AlF_3(4)$ glass rod with a Teflon FEP cladding (Mitachi *et al* 1983b). The minimum loss is 82 dB km^{-1} at a wavelength of 2.6 μm. The transmission loss decreases with increasing wavelength in the shorter wavelength region and increases steeply with wavelength in the longer wavelength range. The former is due to metal impurities, and will be described in detail in the following. The latter arises predominantly from the infrared vibrational absorption of the glass. Several peaks appear in the measured wavelength region. The large peak centered at 2.9 μm is caused by the fundamental stretching vibration of OH ions. The small hump at 4.1 μm originates from the absorption by Teflon FEP cladding. This absorption can be eliminated by changing the cladding material from Teflon FEP to fluoride glass. Furthermore, the loss at 2.9 μm depends largely on the surface condition of the core, which is attacked by the atmospheric H$_2$O. Therefore, it is difficult to reduce the transmission loss of the fluoride glass fiber with Teflon FEP cladding.

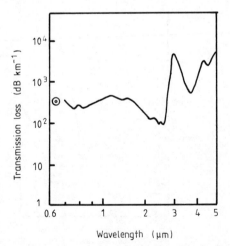

Figure 4.24 The transmission loss spectrum for Teflon FEP clad fiber ($\odot$, He–Ne laser, 0.633 μm) (after Mitachi *et al* 1983b).

The influence of external circumstances on the transmission loss of core-clad type glass fiber can however basically be eliminated. The glass systems studied extensively to date are ZrF_4–BaF_2–GdF_3 and ZrF_4–BaF_2–LaF_3 glasses. The transmission loss of the former is presented in figure 4.25 (Mitachi *et al* 1984). The core glass composition is $ZrF_4(60.5)$–$BaF_2(31.7)$–$GdF_3(3.8)$–$AlF_3(4.0)$ and the cladding composition is $ZrF_4(59.2)$–$BaF_2(31.0)$–$GdF_3(3.8)$–$AlF_3(6.0)$. The fiber was drawn from the preform prepared by the built-in casting method. A smooth core–cladding boundary, which is formed by the built-in casting technique, leads to a reduction in optical loss. The profile of the spectral loss of this fiber is very similar to that for the Teflon FEP clad glass fiber, except that the small hump centered at 4.1 μm, which is observed in the Teflon FEP clad fiber (figure 4.24), disappears. This is due to the fact that the transmitted light power is sufficiently confined to the fiber core. The attenuation in the short wavelength range of less than 2.5 μm is caused by impurity metal ions, particularly Fe^{2+}, Cu^{2+}, Ni^{2+} and Cr^{3+}. The amount of the impurities is estimated to be approximately 0.1–1 ppm. It can be understood that in order to attain an ultra-low loss value, impurity concentrations in fluoride glass fibers should be reduced to a 1 ppb level or less. The small peaks at 2.25 and 2.45 μm are due to combined overtones between the OH stretching vibration and glass matrix vibration of Zr–F and Ba–F bonds, respectively. The minimum loss is 21 dB km^{-1} at 2.55 μm (Mitachi *et al* 1983b).

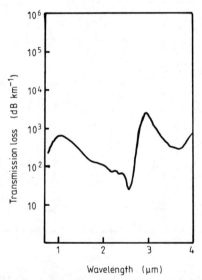

Figure 4.25 The transmission loss spectrum for ZrF_4–BaF_2–GdF_3 based glass fibers (after Mitachi *et al* 1984, © 1984 IEEE).

The lowest reported transmission loss of ZrF_4–BaF_2–LaF_3 based fluoride glass fibers is 0.7 dB km^{-1} (Kanamori and Sakaguchi 1986). The transmission loss spectrum for the fiber is shown in figure 4.26. The composition of the core is ZrF_4–BaF_2–LaF_3–YF_3–AlF_3–NaF, and of the cladding ZrF_4–HfF_4–BaF_2–LaF_3–YF_3–AlF_3–NaF. This fiber exhibits a loss of about 0.7 dB km^{-1} at 2.63 μm, as shown in the figure. The scattering loss is reduced to a value of the order of 0.1 dB km^{-1}. This is due to the higher glass forming ability of glasses containing NaF. Note that the transmission loss is the same order of magnitude as the loss of 0.15 dB km^{-1} for low loss silica glass fiber.

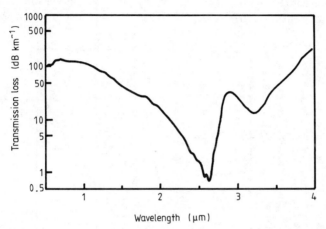

Figure 4.26 The transmission loss spectrum for ZrF_4–BaF_2–LaF_3 based glass fiber (after Kanamori and Sakaguchi 1986).

Besides the above-mentioned compositions of the ZrF_4-based glasses, a variety of glass compositions have been studied for the fabrication of fiber. Table 4.6 lists the compositions and some properties of the fabricated fibers.

The optical loss spectra of HfF_4-based glasses have also been tested. Here the multiphonon absorption edge is located in a longer wavelength region than in ZrF_4-based glasses (Drexhage *et al* 1982a). However, the details have not yet been reported.

From the structural point of view, all the above-mentioned optical fibers are so-called multimode optical fibers. However, single-mode optical fibers also have an important role for long haul optical communication because of their large capacity for light transmission. Ohishi *et al* (1984b) fabricated ZrF_4-based single-mode fibers and measured the transmission loss. The single-mode fiber is prepared using a preform fabricated by the built-in casting method. The preform is jacketed by a fluoride glass tube fabricated by the mold upsetting method, which is similar to built-in casting. This jacketing process is useful for obtaining glass fibers with a core diameter small enough for a single-mode waveguide. The core and cladding diameters are 18 and 92 μm, respectively, and the fiber diameter is 170 μm. The core material composition is $ZrF_4(60.2)-BaF_2(31.2)-GdF_3(3.8)-AlF_3(3.8)-PbF_2(1)$, and the cladding composition is $ZrF_4(59.2)-BaF_2(31.1)-GdF_3(3.8)-AlF_3(5.7)$. The jacketing composition is $ZrF_4(60.5)-BaF_2(31.7)-GdF_3(3.9)-AlF_3(3.9)$. The refractive index difference between the core and cladding is 0.3%. Therefore, the normalized frequency at a 3.28 μm wavelength is estimated to be 2.03, which leads to single-mode operation. The transmission loss for this single-mode optical fiber is shown in figure 4.27. Minimum transmission loss is 160 dB km^{-1} at a wavelength of 3.28 μm.

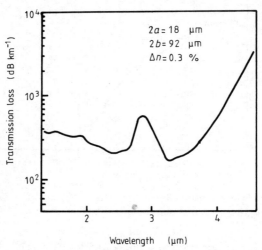

Figure 4.27 The transmission loss spectrum of a fluoride glass single-mode optical fiber ($2a = 18$ μm (core diameter); $2b = 92$ μm (cladding diameter); $\Delta n = 0.3\%$ (refractive index difference) (after Ohishi *et al* 1984b, © 1984 IEEE).

Infrared absorption edge. The infrared absorption edge in fluoride glass plays an important role in predicting an achievable minimum transmission loss in the wavelengths of 2–3 μm. Several reports on the infrared absorption edge have been published to date (Bendow *et al* 1981a, Moynihan *et al* 1981, Drexhage *et al* 1982b, Matecki *et al* 1983).

In general, the dependence of the edge properties on composition can be described in the manner given in section 2.3.2. Large reduced mass and small force constant lead to a longer absorption edge. For example, the infrared edge of the HfF_4-based glass is located at a longer wavelength than in ZrF_4-based glass. AlF_3 is a constituent that is beneficial for glass formation but is undesirable for the infrared edge shift toward the longer wavelength. Various types of the BaF_2/ThF_4

glasses possess the steepest infrared edges at the longest wavelength.

As was pointed out by Tran *et al* (1984b), it is likely that other compositions with even longer wavelength infrared edges exist, but the extent of the possible shift remains uncertain at this time. For example, mixed halide glasses based on CdF_2 appear to have extended transparency at longer wavelengths than both fluorozirconates and rare-earth fluoride glasses, as shown in figure 4.28 (Matecki *et al* 1983). The figure also shows the absorption edges of ZrF_4-, HfF_4- and BaF_2/ThF_4-based glasses for comparison. Appropriate mixed halides could extend transparency while retaining some of the superior physical and chemical properties of fluorides compared to heavier anion halides.

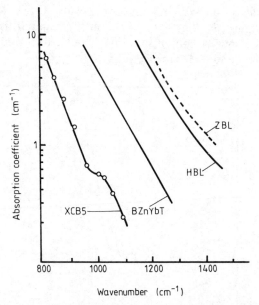

Figure 4.28 Absorption coefficient versus wavenumber for fluorozirconate (ZBL), fluorohafnate (HBL), BaF_2/ThF_4-based (BZnYbT), and CdF_2-based (XCB5) glasses. The XCB5 glass composition is $CdF_2(65)$–$CdCl_2(2)$–$BaCl_2(33)$ (after Matecki *et al* 1983).

The infrared absorption edge can be sensitive to the preparation procedure, although this sensitivity varies considerably with composition. In particular, conditions conducive to the incorporation of oxide impurities in the glass lead to increased edge absorption. For example, improvements in edge characteristics achieved by use of reactive atmospheres have been reported.

Impurity absorption. Impurities may be introduced into the glass at various stages of fiber fabrication. For example, starting materials, melt containers and atmospheres during the melting process are the main sources of the impurities in a fiber. These impurities may influence the absorption edge characteristics and introduce the absorption bands in the interesting wavelength regions. The main types of impurities which influence fluoride glass transparency are the 3d transition metal and rare-earth elements, and the hydroxides. As noted by Tran *et al* (1984b), another important distinction for bulk optical components is whether the impurities are on the surface or in the bulk. Surface impurities in particular are introduced from attack by the atmosphere.

Impurity absorption behaviour was first analyzed systematically by Mitachi *et al* (1983c). Figure 4.29 shows the transmission loss spectrum and its loss factor analysis for the ZrF_4–BaF_2–GdF_3–AlF_3 glass. As shown in the figure, Fe^{2+}, Cu^{2+}, Ni^{2+}, and Cr^{3+} impurities are present. The estimated impurity contents are 0.04, 0.25, 0.09 and 0.11 ppm for Fe^{2+}, Cu^{2+}, Ni^{2+} and Cr^{3+}, respectively.

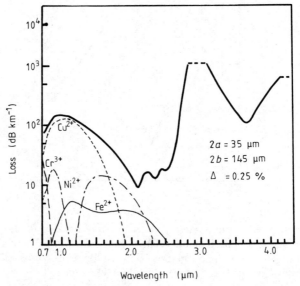

Figure 4.29 The transmission loss spectrum for fluoride-based optical fiber and its loss factor analysis (after Mitachi *et al* 1983c).

Ohishi *et al* (1983) investigated the detailed impurity behaviors. Absorption losses associated with given concentrations of impurities have been determined at selected wavelengths, as shown in table 4.8. Figures 4.30 and 4.31 show the calculated absorption loss spectra for the most common contaminants. The absorptivities of Fe, Ni, Co and Cu decrease rapidly above 2 μm, and thus their effect on absorption losses in fluoride glass fibers, whose transmission window lies between 2.0 and 5.0 μm, is less severe than in silica fibers. Ce, Pr, Nd, Sm, Eu, Tb and Dy, on the other hand, exhibit sharp absorption bands above 2 μm and are detrimental to fluoride glass fiber materials containing LaF_3 or GdF_3. The impurity level required to produce a 0.1 dB km^{-1} loss at 2.5 μm has been calculated to be 3.6 ppb for Fe, 3.2 for Co, 17 for Ni, 714 for Cu, 57 for Pr, 5 for Nd, 38 for Sm, 71 for Eu, and 143 for Dy (Ohishi *et al* 1983).

Table 4.8 Absorptivities (dB km^{-1}/parts in 10^6) of transition metals and rare earths at different wavelengths (after Ohishi *et al* 1983).

Element	λ (μm)			
	2.0	2.5	3.0	3.5
Fe	90	28	2	0.9
Co	130	31	4	0.75
Ni	90	6	0.5	0.05
Cu	3	0.14	0.01	–
Ce	–	–	7.7	50
Pr	27	1.8	0.4	4.5
Nd	–	20	0.6	–
Sm	4.4	2.6	5.6	1.2
Eu	12	1.4	3.1	14.3
Tb	25	–	14.3	1.1
Dy	0.06	0.7	8.3	1.0

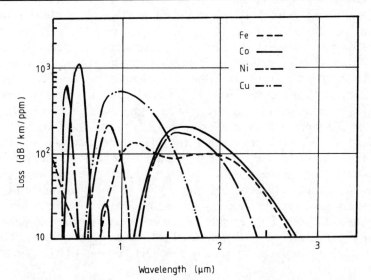

Figure 4.30 Calculated absorption loss spectra due to 1 part in 10^6 of transition metals (after Ohishi *et al* 1983).

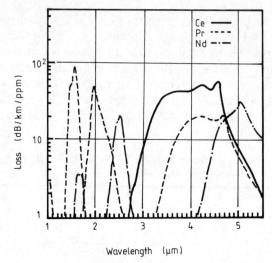

Figure 4.31 Calculated absorption loss spectra due to 1 part in 10^6 of rare-earths (after Ohishi *et al* 1983).

A primary absorption in the mid-infrared is associated with OH, which can enter the glass during the batching and/or the melting process. Reactive atmospheres such as CCl_4, SF_6 and NF_3 are very useful for reducing the number of OH ions (Robinson *et al* 1980, Tran *et al* 1984b, Nakai *et al* 1986b). Figure 4.32 shows the difference between the loss spectrum of a glass after SF_6 reactive atmosphere processing and that of a non-treated glass (Tran *et al* 1984b). In the figure, an intensive absorption peak in spectrum A corresponds to $78\,000$ dB km^{-1} loss at $2.9\,\mu$m due to surface and bulk OH for a ZrF_4-based glass processed under an inert Ar atmosphere and quenched in air; the effect of processing the same sample under SF_6 is shown in spectrum B. The residual OH absorption band in curve B remains unchanged with sample thickness, which suggests that essentially only surface OH remains after such processing.

Maze *et al* (1984) reported that low OH content fluoride bulk glasses and fibers can also be obtained by simply refining the fluoride glass melts at a high temperature of $\sim 1000\,°$C under a neutral atmosphere, since the hydroxyl groups are not stable in the glass melt at high temperatures. They obtained a total loss of 110 dB km^{-1} at $2.9\,\mu$m in a fluoride glass fiber. Furthermore, under moderate melting temperatures but dry and neutral glove box atmosphere processing conditions, Tran *et al* (1984b) obtained an extremely low OH content of 50 dB km^{-1} at $2.9\,\mu$m.

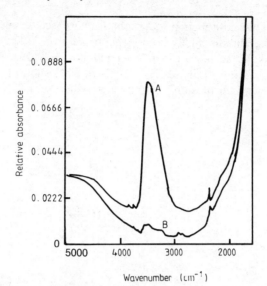

Wavenumber (cm^{-1})

Figure 4.32 OH absorption of a $ZrF_4(51.3)$–$BaF_2(20.47)$–LaF_3 (5.27)–$AlF_3(3.24)$–$LiF(19.49)$ glass before and after SF_6 reactive atmosphere processing (after Tran *et al* 1984b).

Poulain and Saad (1984) studied the influence of $NaPO_3$, $KHSO_4$, Na_2CO_3, $NaNO_3$, and $TiOF_2$ on the absorption loss. They found that complex anions such as sulfate and phosphate may have a drastic influence on the optical absorption of fluorozirconate, and the acceptable residual level is well below 1 ppm. They also found that other impurities such as carbonate and nitrate are not stable in bulk glass and may therefore be conveniently removed.

Scattering loss. It is important to measure scattering losses, because losses such as Rayleigh scattering loss fix the limitation of the transmission loss. Ohishi *et al* (1982) investigated the light scattering mechanism of the ZrF_4-based glass. They found that light scattering depends strongly on the melt cooling rate through the crystallization temperature.

Scattering loss can basically be represented as the sum of intrinsic scattering losses, such as Rayleigh scattering, and the extrinsic scattering losses due to defects in the glass. Sakaguchi and Takahashi (1986) studied extensively the behavior of light scattering in fluoride glass fibers. Figure 4.33 shows the intensity distribution of light scattering from a small homogeneous region. This scattering corresponds to Rayleigh scattering. Silica glass exhibits a cocoon-like distribution, which means typical Rayleigh scattering. ZrF_4–BaF_2–GdF_3–AlF_3 glass behaves like a silica glass although some slight forward scattering occurs. This means that the glass is highly homogeneous. On the other hand, glasses containing alkalis such as Li and Na exhibit backward scattering. In particular, ZrF_4–BaF_2–LaF_3–YF_3–AlF_3–NaF glass shows large backward scattering, which indicates an inhomogeneous nature.

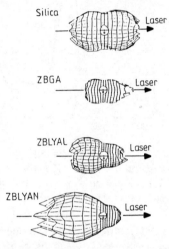

Figure 4.33 Intensity distributions for light scattering from the small homogeneous regions in various fluoride glass fibers (ZBGA: ZrF_4–BaF_2–GdF_3–AlF_3; ZBYAL: ZrF_4–BaF_2–LaF_3–YF_3–AlF_3–LiF; ZBLYAN: ZrF_4–BaF_2–LaF_3–YF_3–AlF_3–NaF) (after Sakaguchi and Takahashi 1986).

Figure 4.34 shows the wavelength dependence of the scattering loss originating from the homogeneous region. As shown, the scattering loss is proportional to $(\text{wavelength})^{-4}$. In ZrF_4–BaF_2–GdF_3–AlF_3 glass, which exhibits a cocoon-like scattering, the gradient of the curve is $0.6\,\text{dB}\,\text{km}^{-1}\,\mu\text{m}^4$. This value is smaller than that of silica glass $(1\,\text{dB}\,\text{km}^{-1}\,\mu\text{m}^4)$. Furthermore, the gradient of ZrF_4–BaF_2–LaF_3–YF_3–AlF_3–LiF glass is $0.7\,\text{dB}\,\text{km}^{-1}\,\mu\text{m}^4$, which is also smaller than for silica glass. This shows that Rayleigh scattering in these glasses is small compared to silica glass. Sakaguchi and Takahashi (1986) estimated from these results that the minimum losses of these fluoride glass fibers are below $0.01\,\text{dB}\,\text{km}^{-1}$ at around $2.5\,\mu\text{m}$. Loss in ZrF_4–BaF_2–LaF_3–YF_3–AlF_3–NaF glass, on the other hand, shows a large wavelength dependence. This means that another origin of scattering, such as phase separation, exists in the glass.

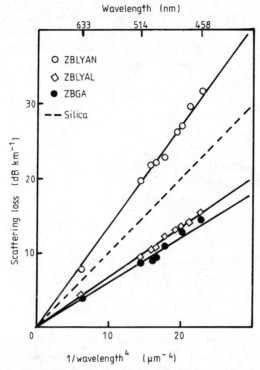

Figure 4.34 Wavelength dependences of the scattering losses originating from the homogeneous regions. The notation used is the same as in figure 4.33 (after Sakaguchi and Takahashi 1986).

Bright scattering centers causing scattering loss due to defects, however, behave differently from the homogeneous scattering. Figure 4.35 shows the scattering light distribution of a typical scattering center. Strong forward scattering can be seen, which would be the origin of Mie scattering. The size of this center can be estimated to be of the submicron order. These scattering centers are thought to be microcrystals, oxides, pores, impurity particles, segregations, or irregularities between core and cladding. In the present stage of research, these scatterings cause large transmission loss in the fiber, and so transmission loss could be reduced further by careful elimination of these scattering centers.

Tokiwa *et al* (1985c) studied the relation between the scattering intensity and the reheated condition in fluorozirconate glass to clarify the effect of devitrification induced by fiber drawing using the preform method. Drawing-induced scattering is substantially caused by the nucleation of microcrystallites. Therefore, it was pointed out that optimized glass compositions and reheating conditions can allow fiber drawing without increase of scattering loss. Tokiwa *et al* (1985b) insisted that a ZrF_4–BaF_2–NaF–LaF_3–AlF_3 core, ZrF_4–BaF_2–NaF–LaF_3–AlF_3–HfF_4 clad glass fiber is preferable, with transmission loss expected to be 0.0045 dB km^{-1} at a 4 μm wavelength.

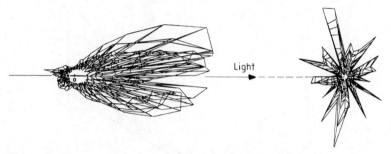

Figure 4.35 The scattered light distribution for a high scattering fiber segment (after Sakaguchi and Takahashi 1986).

Furthermore, it should be noted that the scattering loss is closely related to oxygen impurities in fluoride glass (Nakai *et al* 1985a, Mitachi *et al* 1985). Nakai *et al* (1985a) pointed out that scattering centers are already present in the glass melt before fiber drawing. Therefore, their number can be considerably reduced by melting the glass for a longer period. However, a detailed creation mechanism for scattering centers has not yet been clarified.

Ohsawa and Shibata (1984) reported that the addition of sulfate to fluoride glass is effective in reducing light scattering, as is the addition of SO_3F^-, BF_4^- and InF_3.

4.2.4.2 Refractive index and dispersion

The refractive indices of most fluoride glasses lie in the range 1.47–1.53, which is comparable to the conventional silica-based glasses. It is well known that the refractive index can be changed with dopant concentrations. Figure 4.36 shows the relation between dopant concentration and refractive index in the ZrF_4–BaF_2–GdF_3 glass system (Miyashita and Manabe 1982). As shown, heavier and more polarizable fluorides such as PbF_2 increase the index, while lighter and less polarized ones such as LiF decrease it. On the other hand, addition of YF_3 up to 8 mol.% does not cause any change in refractive index (Mitachi *et al* 1983b).

It is very important for an optical fiber with a high bandwidth to operate in single-mode condition at the wavelength where the material dispersion is zero. In order to determine the wavelength for the zero material dispersion, the wavelength dependence of the refractive index

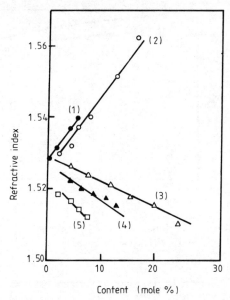

Figure 4.36 The relation between dopant concentration and refractive index in the ZrF_4–BaF_2–GdF_3 glass system, for (1) PbF_2, (2) BiF_3, (3) LiF, (4) NaF, (5) AlF_3 (after Miyashita and Manabe 1982, © 1982 IEEE).

must be measured. Figure 4.37 shows the refractive indices for various wavelengths for $ZrF_4(62)$–$BaF_2(33)$–$LaF_3(5)$ and $HfF_4(62)$–$BaF_2(33)$–$LaF_3(5)$ glasses (Bendow *et al* 1981b). The refractive index for silica glass is also shown for comparison. The refractive index of fluoride glass varies relatively slowly with the wavelength. Figure 4.38 shows the relations between refractive index and wavelength for $ZrF_4(63)$–$BaF_2(33)$–$GdF_3(4)$ and $AlF_3(40)$–$BaF_2(22)$–$CaF_2(22)$–$YF_3(16)$ glasses (Miyashita and Manabe 1982). In addition, figure 4.39 shows refractive index spectra for AlF_3-doped ZrF_4-based glasses (Miyashita and Manabe 1982).

The material dispersion, which is proportional to the value $d^2n/d\lambda^2$, can then be calculated by using the wavelength dependence of the refractive index n. Figure 4.40 shows the material dispersion of $ZrF_4(62)$–$BaF_2(33)$–$LaF_3(5)$ and $HfF_4(62)$–$BaF_2(33)$–$LaF_3(5)$ glasses. The figure shows that zero dispersion arises in the region of 1.6–1.7 μm, whereas the loss minimum occurs in the 3–4 μm region indicated in the previous section. However, it must be noted that the magnitude of the material dispersion is small enough through an extended range of wavelengths which includes the minimum loss region. Table 4.9 shows the wavelengths of zero material dispersion for various fluoride glasses (Mitachi *et al* 1983b). As shown in the table, Pb-doped fluoride glass gives the longest wavelength of 1.704 μm.

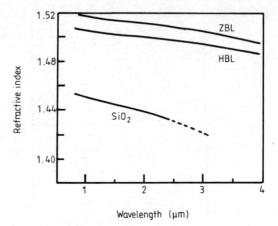

Wavelength (μm)

Figure 4.37 Refractive index versus wavelength for $ZrF_4(62)$–$BaF_2(33)$–$LaF_3(5)$ and $HfF_4(62)$–$BaF_2(33)$–$LaF_3(5)$ glasses (after Bendow *et al* 1981b).

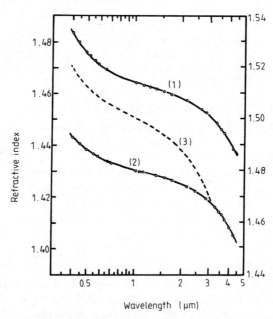

Wavelength (μm)

Figure 4.38 Refractive index spectra for ZrF_4-based glasses, for (1) $ZrF_4(63)$–$BaF_2(33)$–$GdF_3(4)$, (2) $AlF_3(40)$–$BaF_2(22)$–$CaF_2(22)$–$YF_3(16)$, (3) SiO_2 (after Miyashita and Manabe 1982, © 1982 IEEE).

The material dispersion for ZrF_4–BaF_2–LaF_3–AlF_3 glass has also been studied by Brown and Hutta (1985). It was found that the zero material dispersion wavelength is 1.62 μm and that the dispersion properties are not enhanced by the addition of NaF. The material dispersion can also

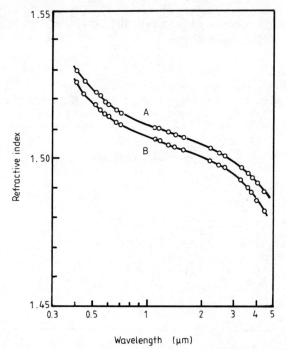

Figure 4.39 Refractive index spectra for AlF_3-doped ZrF_4-based glasses, for A, 2 mol.% AlF_3-doped glass; B, 6 mol.% AlF_3-doped glass (after Miyashita and Manabe 1982, © 1982 IEEE).

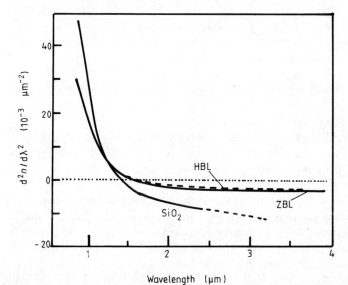

Figure 4.40 Material dispersions of $ZrF_4(62)$–$BaF_2(33)$–$LaF_3(5)$ and $HfF_4(62)$–$BaF_2(33)$–$LaF_3(5)$ glasses compared to fused silica (after Bendow *et al* 1981b).

be found directly using a single-mode fiber from chromatic dispersion measurements (Monerie *et al* 1985).

Table 4.9 Wavelengths giving zero material dispersion for various fluoride glasses. BGZA is BaF_2–GdF_3–ZrF_4–AlF_3 (after Mitachi *et al* 1983b).

Glass	Zero material dispersion wavelength (μm)
BGZA (2 mol.%)	1.682
BGZA (4)	1.675
BGZA (6)	1.670
BGZA–Li (4)	1.760
BGZA–Cs (4)	1.668
BGZA–Cd (4)	1.681
BGZA–Sn (4)	1.683
BGZA–Pb (4)	1.704
BGZA–Y (4)	1.673

It is however possible to exploit waveguide dispersion in optical fibers to compensate for material dispersion (Gannon and Byron 1982). For example, the optimum core size would appear to be around 12 μm, which would give zero total dispersion at a 2.5 μm wavelength. At the lower loss wavelength of 3 μm, a total dispersion of about 3 ps $\text{Å}^{-1}\,\text{km}^{-1}$ would be estimated. This would allow a system bandwidth of beyond 300 GHz $\text{Å}^{-1}\,\text{km}^{-1}$. Thus, fluoride glasses offer the possibility of very large information carrying bandwidths over a wide spectral range.

4.2.4.3 Mechanical and thermal properties

The mechanical strength of fluoride glasses is, in general, poor, and their thermal stability is lower than that of silica glasses. However, no serious problems arise if mechanical and thermal attacks from the environment are carefully eliminated.

In table 4.10, typical values of density, thermal expansion coefficient, glass transition temperature T_g, and crystallization temperature T_x for representative fluoride glasses are shown, with the comparisons of fused silicas and chalcogenides (Drexhage *et al* 1980). Typical values of glass transition temperature T_g are in the range 300–325 °C for fluorozirconates, 350–450 °C for BaF_2/ThF_4 glasses, and less than 300 °C for glasses highly doped with alkalis. Typical densities are 4.5–5 g cm^{-3} for fluorozirconates, and 5.5–6 g cm^{-3} for fluorohafnates and BaF_2/ThF_4 glasses.

It is very important in fiber drawing to make $\Delta T = T_x - T_g$ large for the wide working range. In general, ΔT of the fluoride glasses is in the range 70–100 °C, which is smaller than for conventional glass formers (Tran *et al* 1984b).

Table 4.10 Selected properties of fluoride glasses (after Drexhage *et al* 1980).

Glass	Refractive index (n_D)	Abbe no. v	Density (g cm^{-3})	Thermal expansion (10^{-7}°C^{-1})	Transition temp. T_g(°C)	Crystalliz. temp. T_x(°C)
HBL	1.522	–	5.87	191.2	315	389
HBLC	1.506	–	5.82	174.8	306	391
HBLPC	1.525	75	6.12	–	–	–
HBLAPC	1.526	–	6.18	–	–	–
HBTKLRC	1.504	72	5.87	–	312	389
ZBTNLRC	1.517	86	4.68	–	–	–
HBTKLLi	1.508	–	5.90	155.7	–	–
ZBTLLi	1.518	–	4.66	213.6	–	–
Fused silica	1.458	68	2.20	5.5	1100	~1575
As$_2$S$_3$	2.403 (2.5 μm)	–	3.43	237.0	208	–
As$_2$Se$_3$	2.769 (3.0 μm)	–	4.75	210.0	187	–

H = hafnium, Z = zirconium, B = barium, L = lanthanum, Li = lithium, C = cesium, T = thorium, K = potassium, P = lead, N = sodium, A = aluminum, R = rubidium

The hardness and fracture toughness of fluoride glasses are generally lower than those of high silica glasses, but higher than those of chalcogenide glasses, as shown in table 4.11 (Bendow 1984). Measured rupture strengths of bulk specimens are reported to be up to 35 klb in^{-2}, but these values are a reflection of their surface condition rather than an indicator of the ultimate strength of the material. The fracture toughness, on the other hand, is an appropriate measure of intrinsic strength, and the values given in table 4.11 imply strengths for fluoride glasses that are roughly one third of those of high silica glasses (Tran *et al* 1984b).

The tensile strength of the fluorozirconate glass fiber is usually 60–220 MPa, which depends on the fiber drawing temperature (Lau *et al* 1985). At a high drawing temperature, microcrystal growth occurs in the fiber which degrades the fiber strength. When the drawing temperature is sufficiently low, fiber strength in the range 500–700 MPa can be observed. Although these values are still only one tenth of the estimated theoretical strength, the intrinsic strength of fluoride fibers is unlikely to pose problems from the viewpoint of application.

It should be noted that the surface quality of the preform before fiber drawing plays an important role in improving the fiber strength. Surface treatments by appropriate methods are therefore necessary if high strength fibers are to be obtained (Schneider *et al* 1986).

Table 4.11 Mechanical properties of fluorozirconate glasses (V, Vickers hardness; K_p, Knoop hardness) (after Bendow 1984).

Material	Hardness (kg mm^{-2})	Fracture toughness (MPa m$^{1/2}$)	Fracture strength (MPa)	Thermal expansion (10^{-7} °C^{-1})
Fluorozirconate type	225–250 (V)	0.25–0.27	20	150–180
Chalcogenide	100–200 (V)	0.20	20	240–250
Fused silica	~800 (V)	0.7–0.8	70	5.5
Calcium fluoride	~120–160 (K_p)	0.35	(40)	~180

4.2.5 Summary

Fluoride glass fibers offer the best prospect of ultra-low loss optical fibers whose transmission loss is lower than that of the high silica glass fiber. Fluoride glass fibers have various desirable properties, such as a broad transparency range spanning the mid-infrared to near-infrared, low refractive index and dispersion, and low Rayleigh scattering.

The fluoride glasses studied so far are mainly fluorozirconate and fluorohafnate glasses. These exhibit a lesser tendency toward crystallization and higher infrared transparency. ZrF_4 (or HfF_4) based glasses contain ZrF_4 (or HfF_4) as a glass network former (50–70 mol.%), BaF_2 as a primary network modifier (about 30 mol.%), and one or more additional metal fluorides of the rare-earths, alkalis, or actinides as glass stabilizers.

In order to fabricate low loss glass fibers, the loss factors occurring at each stage of the fabrication process must be eliminated. In particular, purification of the starting materials and fiber drawing are the most important processes. Reactive atmosphere processing or the sublimation method are proposed for purification. Built-in casting, rotational casting, or reactive vapor transport methods have also been proposed for the fabrication of the preform. After several recent extensive studies a transmission loss of less than 1 dB km^{-1} has been obtained (Tran 1986, Sakaguchi and Takahashi 1986). However, further extensive studies are definitely required for reducing both impurity absorption and scattering loss.

4.3 Chalcogenide glass fibers

4.3.1 Introduction
Chalcogenide glasses are defined as materials containing at least one of the elements S, Se and Te. They are available in a stable vitreous state and with a wide optical transmission range.

In the early stages of the research on infrared optical fibers, Kapany and Simms (1965) prepared As–S glass optical fibers which showed a relatively high optical loss of 20 dB m^{-1} at 5.5 μm wavelength. At first, chalcogenide glass fibers were studied mainly for applications such as image transmission, and so transmission loss did not merit much attention. However, as the optical communication research progressed, loss reduction became the main target of research on chalcogenide glass optical fibers.

In this section materials, fabrication methods and properties of chalcogenide glass optical fibers are described.

4.3.2 Materials
4.3.2.1 Sulfide glasses
Sulfide glasses are, in principle, divided into two groups: arsenic–sulfur glasses and germanium–sulfur glasses.

The arsenic–sulfur glasses are one of the most practical choices of chalcogenide glasses for infrared fiber optics. Variations of arsenic–sulfur glass systems can be easily obtained, which means that the fabrication of core-clad type optical fibers with suitable numerical apertures is possible. This glass has a softening point of 205 °C and a thermal expansion coefficient of 250×10^{-7}°C^{-1}. It gives a relatively broad range of transmission between 0.6 and 10 μm (when 10 mm thick) and has a relatively high refractive index of 2.41. Figure 4.41 shows the glass forming region of the ternary As–Ge–S glass system, which is the typical As–S based glass system.

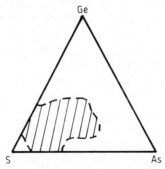

Figure 4.41 The glass forming region of a Ge–As–S system (after Savage 1965).

From the historical point of view, Kapany and Simms (1965) first fabricated the As_2S_3 glass optical fibers. However, their transmission losses were relatively high, around $20 \, dB \, m^{-1}$ at a $5.5 \, \mu m$ wavelength. Recently, unclad fibers were drawn from an As_2S_3 rod (Miyashita and Terunuma 1982). Figure 4.42 shows composition ranges which can be drawn into fibers by preform and crucible techniques. Various As–S glass optical fibers were also fabricated. Table 4.12 lists the As–S glass systems examined so far.

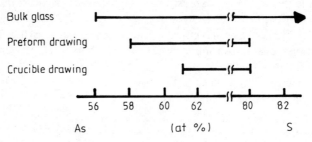

Figure 4.42 Fiber drawing regions in an As–S system for the preform and crucible methods (after Kanamori *et al* 1984, © 1984 IEEE).

Germanium–sulfur glass systems, on the other hand, have a relatively high softening temperature, large glass forming area, and low toxicity. Shibata *et al* (1980b, 1981a,b) have studied them extensively. Figure 4.43 shows glass forming and fiber drawing compositions for the Ge–P–S glass system. The latter is smaller than the former because of devitrification in the Ge-rich region and sublimation of sulfur in the S-rich region at a temperature above the softening point. It was reported that the appropriate glass composition for GeS_x ranges from $x = 2$ to 4. It was also reported that the addition of phosphorous to the germanium sulfide glass increases glass forming ability and optical homogeneity if the amount of phosphorous is less than 10 mol.%.

Table 4.13 shows the physical properties of Ge–S glasses. The optical transmission range for Ge–S glass is almost the same as that for As–S glass, giving a broad transmission between 0.5 and 11 μm (when 2 mm thick). Ge–S glasses have a relatively low refractive index (n_D) of about 2.1.

4.3.2.2 Selenide glasses

Various selenide glass systems have been studied mainly in order to achieve lower loss at the wavelengths of 5.4 μm (CO laser) and 10.6 μm (CO_2 laser). Selenide glasses are divided into two groups, one based on

Table 4.12 As–S chalcogenide glass systems studied so far.

Material	Structure	Property	Fabrication	Reference
As_2S_3	Unclad, clad	20 dB m^{-1} (5.5 μm)	Preform, crucible	Kapany and Simms (1965)
As–S	–	8–10 dB m^{-1} (5.3 μm)	–	Vechkanov et al (1982)
As–S	Unclad	0.078 dB m^{-1} (2.4 μm)	Preform	Miyashita and Terunuma (1982)
As–S	Unclad, polymer clad	~1 dB m^{-1} (~4 μm)	Preform	Vasiliev et al (1983)
As–S	Clad	0.035 dB m^{-1} (2.44 μm)	Double crucible	Kanamori et al (1984)
As–S	–	0.3 dB m^{-1} (5.3 μm)	Preform	Hattori et al (1984)
As–S	–	~1 dB m^{-1} (~5 μm)	–	Andriesh et al (1984)
As_2S_3	Unclad	0.17 dB m^{-1} (5.4 μm)	Preform	Arai and Kikuchi (1984)
$As_{40}S_{60}$	Unclad	0.064 dB m^{-1} (2.4 μm)	Preform	Kanamori et al (1985)
As–S	Image guide	0.6 dB m^{-1} (3.8, 4.8 μm)	Preform	Saito et al (1985)
As–S	Teflon clad	0.15 dB m^{-1} (4.8 μm)	Preform	Saito and Takizawa (1986)

As–Se glass and the other on Ge–Se glass. This classification method is just same as in the case of sulfide glass systems. The As–Se glass systems include As–Se and As–Ge–Se (As rich) glasses. Ge–Se glasses so far studied are Ge–Se, Ge–As–Se (Ge rich), La–Ga–Ge–Se and Ge–Sb–Se glasses. Glass forming regions of the typical selenide glass systems Ge–Se, Ge–As–Se and Ge–Sb–Se are shown in figure 4.44. Table 4.14 lists the selenide glass systems studied so far for the fabrication of optical fibers.

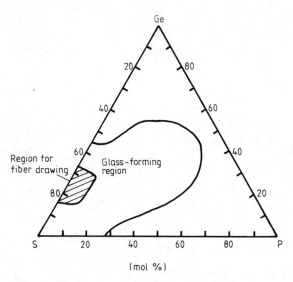

Figure 4.43 Appropriate glass composition for fiber drawing (shaded region). The area surrounded by the solid line shows the glass forming region (after Shibata *et al* 1981b).

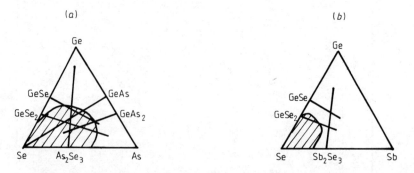

Figure 4.44 Glass forming regions (shaded) of typical selenide glass systems: (*a*) Ge–As–Se, (*b*) Ge–Sb–Se (after Hilton *et al* 1966).

Table 4.13 Ge–S chalcogenide glass systems studied so far.

Material	Structure	Fabrication	Property	Reference
$Ge_3P\,S_{7.5}$	Unclad	Preform	0.4 dB m^{-1} (2.5 μm)	Shibata et al (1980b)
$Ge\,S_3$	Unclad, plastic clad	–	<1 dB m^{-1}	Shibata et al (1981a)
$Ge\,S_3$	Teflon FEP clad	Preform	<1 dB m^{-1} (2.4, 3.3, 4.7, 5.1 μm)	Shibata et al (1981b)

Table 4.14 Selenide glass systems studied so far for the fabrication of optical fibers.

Material	Structure	Fabrication	Property	Reference
As_2Se_3	–	–	0.06 dB m^{-1} (5.56 μm)	Vlasov et al (1982)
As_2Se_3	As_2Se_3 core As_2S_3 clad	Preform	8–10 dB m^{-1} (5.3 μm)	Vechkanov et al (1982)
$As_{38}Ge_5Se_{57}$	Unclad and Teflon clad	Preform	0.5 dB m^{-1} (10.6 μm) at LN$_2$ temperature	Takahashi et al (1983)
As_2Se_3	Unclad, As_2S_3 clad	Preform	0.58–0.65 dB m^{-1} (3.8–4.2, 4.5 μm)	Vasiliev et al (1983)
$As_{38}Ge_5Se_{57}$	Unclad	Preform	0.182 dB m^{-1} (2.12 μm)	Kanamori et al (1984)
$As_{38}Ge_5Se_{57}$	Unclad	Preform	0.29 dB m^{-1} (3.40 μm)	Kanamori et al (1985)
$As_{2-x}Se_{3+x}$	Unclad	Crucible, preform	10 dB m^{-1} (10.6 μm)	Bornstein et al (1985)
$Ge_{30}As_{15}Se_{55}$	Unclad	Preform	10 dB m^{-1} (5.5–7.0 μm)	Boniort et al (1980)
La–Ga–Ge–Se				Gannon (1981)
$Se_{55}Ge_{30}As_{15}$	Plastic clad	Preform	10 dB m^{-1} (4–11 μm)	Brehm et al (1982)
$As_{15}Ge_{30}Se_{55}$	Unclad, plastic clad	Preform	10–16 dB m^{-1} (10.6 μm)	Le Sergent (1982)
Ge–Se,	Unclad	Nozzle	0.2 dB m^{-1} (5.5 μm)	Katsuyama et al (1984)
Ge–Sb–Se				
Ge–As–Se	Teflon clad	Preform	0.53 dB m^{-1} (3.8–4.0 μm)	Dianov et al (1984)
Ge–Se	–	Plasma CVD	–	Blanc and Wilson (1985)
Ge–Se	–	CVD	–	Katsuyama et al (1986)

Selenide glasses have, in principle, a wide transparent region compared to the sulfide glasses. For example, a transmission loss of less than $1 \, dB \, m^{-1}$ has been observed in the wavelength region between 2 and $10 \, \mu m$ (Katsuyama *et al* 1984). However, selenide glasses tend generally to have a lower softening temperature than sulfide glasses, so that temperature increases must be avoided when these glass fibers are used for laser power transmission. Nevertheless, selenide glasses are advantageous because they exhibit a stable vitreous state resulting in flexibility of the fiber, and are therefore one of the most interesting candidates for infrared laser power transmission and wide bandwidth infrared light transmission, such as in radiometric thermometers.

4.3.2.3 Telluride glasses

The Se-based chalcogenide glasses described above have relatively wide transparent regions. However, their losses at $10.6 \, \mu m$, which is the wavelength for CO_2 laser light transmission, are still higher than the $1 \, dB \, m^{-1}$ required for practical use. To lower the transmission loss due to the lattice vibration, Te atoms, which are heavier than Se, must be introduced to shift the infrared absorption edge toward a longer wavelength. However, Te-based glasses are difficult to form because of their relatively high tendency to crystallization.

The Te-rich glass optical fibers have been reported by Katsuyama and Matsumura (1986). The transmission losses of their $Ge_{22}Se_{20}Te_{58}$ glass fibers at a $10.6 \, \mu m$ wavelength were reduced down to nearly $1 \, dB \, m^{-1}$. The glass forming region of the Ge–Se–Te glass system is shown in figure 4.45. It can be easily understood that the glass-forming area becomes small with the increase of the Te content. In particular, it is impossible to obtain binary Ge–Te glass when a conventional quenching technique such as water quenching is used.

Other Te-containing glass fibers have been reported by several research institutes. They include $As_{13}Ge_{25}Se_{27}Te_{35}$ glass fiber with a $2 \, dB \, m^{-1}$ loss (Wehr and Le Sergent 1986) and $As_{40}Se_{40}Te_{20}$ glass fiber with a $5 \, dB \, m^{-1}$ loss at $10.6 \, \mu m$ (Pitt *et al* 1986).

4.3.3 Fabrication methods

4.3.3.1 Bulk glass fabrication

In order to fabricate low loss optical fibers, high quality bulk glasses must be obtained. So far, in principle, chalcogenide glasses have been fabricated mainly by two methods. One is based on the sintering of the metal elements in an evacuated ampoule. The other is a chemical vapor deposition method, which has great potential for reducing inhomogeneity of the glass because of its excellent mixing effect on the starting material.

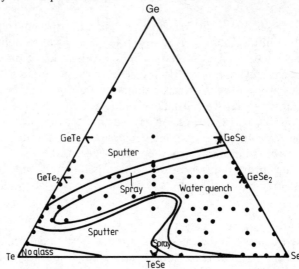

Figure 4.45 Glass formation in the Ge–Se–Te system, showing the quenching techniques required to produce single-phase amorphous samples. Dots correspond to compositions successfully prepared as single-phase glasses (after Sarrach and De Neufville 1976).

Sintering of glass. The sintering method of a binary Ge–Se glass is as follows. Usually, Ge ingots with a purity of more than 99.999 999 99% and Se shots with 99.999% purity are used as the starting materials. A thick quartz tube, closed at one end, is prepared by rinsing its inside with HF and then washing it thoroughly with distilled water before drying in an air oven. Each tube is evacuated to more than 10^{-5} Torr and outgassed by prolonged heating with an oxy-hydrogen torch. After cooling under vacuum, each tube is filled with Ar gas at atmospheric pressure, removed from the vacuum system and filled with the previously weighed Ge and Se, and then repumped to more than 10^{-5} Torr and held at this pressure for about 1 hour before the open end of the tube is sealed with an oxy-hydrogen torch.

Each sealed tube (ampoule) is then loaded into a rocking furnace, as shown in figure 4.46. Details of the rocking furnace are given by Ford and Savage (1976). The furnace temperature is maintained at 800 °C for 35 hours to mix the constituents. After that, the heated ampoule is air quenched in an insulated container at room temperature. Finally, the sintered chalcogenide glass block is obtained by cutting the sealed ampoule.

There is no noticeable difference in sintering the glass even if the starting materials are different from the above-mentioned Ge and Se. However, in general, it should be noted that contamination by impurities, particularly the contamination of the starting materials and silica glass ampoule, must be minimized.

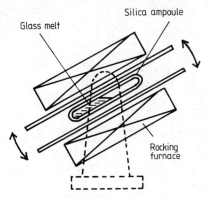

Figure 4.46 Heating of glass ampoules by a rocking furnace.

Reduction of impurity absorption. In a chalcogenide glass, impurity absorptions due to oxygen and hydrogen cause transmission losses in the wavelength region of interest. The reduction of these impurities is therefore important.

It is known (Hilton *et al* 1975) that Al has the effect of reducing the impurity absorption due to oxygen. In order to show the difference between the transmission properties of Al-doped and undoped bulk glass samples, their typical transmission spectra are given in figures 4.47(*a*) and (*b*) (Katsuyama *et al* 1984). Note that the glass samples were not purified so that the effect of Al doping could be studied. In the figure, curve A shows the transmission for the undoped glass block, whose composition is Ge : Se = 20 : 80 mol.%. The absorption bands at the wavelengths of 2.8, 4.5, 6.3, and 12.8 μm are due to OH, Se–H, H_2O, and Ge–O absorptions, respectively. In contrast, the bulk glass doped with 100 ppm of Al shown in curve B exhibits no impurity absorption bands except the 4.5 μm band due to Se–H. The absorption bands of 13.5 μm and those longer than 15 μm are due to intrinsic Ge–Se lattice vibrations. Thus, it is found that adequate Al doping is effective in eliminating OH, H_2O, and Ge–O absorptions. These results can be interpreted in terms of the 'gettering' action of Al (Hilton *et al* 1975). Al is one of the elements which are easily oxidized, so that oxygen atoms can be removed from OH, H_2O, and Ge–O because of the selective combination of oxygen with Al. However, one must take care to avoid additional scattering loss resulting from the Al doping. Figure 4.48 shows scattering loss as a function of Al content for $Ge_{20}Se_{80}$ glass fibers. It can be seen from the figure that the scattering loss increases rapidly for Al doping levels above 100 ppm. This may be due to the increase in scattering centers created by the attack of Al on the quartz ampoule during the melting process. The doping level of Al must therefore not exceed 100 ppm if additional scattering loss is to be avoided.

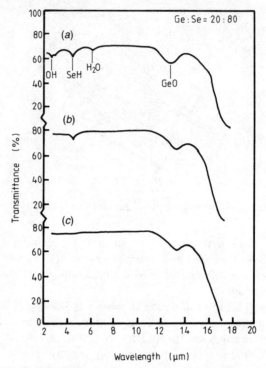

Figure 4.47 Transmission spectra for $Ge_{20}Se_{80}$ glass blocks of 5 mm thickness, for (*a*), undoped; (*b*), Al-doped; (*c*), Al-doped and purified samples (after Katsuyama *et al* 1984).

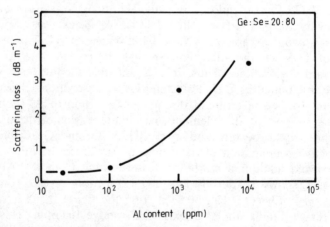

Figure 4.48 Wavelength-independent scattering loss as a function of the Al content for $Ge_{20}Se_{80}$ glass fibers (after Katsuyama *et al* 1984).

Figure 4.47(*b*) shows that the Se–H absorption at 4.5 μm still remains in spite of the fact that other impurity bands vanish completely. The loss reaches as high as $100 \, \text{dB} \, \text{m}^{-1}$. It is found that hydrogen impurities in the Se shots and the water in the silica glass ampoule react with Se during the glass melting process, forming Se–H bonds. In order to eliminate the Se–H bonds, additional purification is required. Se shots can be distilled in a reactive $SeCl_2$ gas atmosphere; the apparatus for this is shown in figure 4.49. Se shots in part A are heated in the $SeCl_2$ gas atmosphere, then the evaporated Se is deposited at part B. The fine Se particles which are deposited are consolidated by heating at 250 °C. Furthermore, ampoules made of water-free silica glass must be used to avoid the reaction of water with Se. The transmission spectrum obtained is shown in figure 4.47(*c*). In the figure, the Se–H absorption cannot be observed.

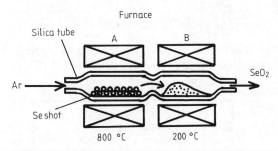

Figure 4.49 Apparatus for Se shot distillation.

Another method for distillation has been reported by Shibata *et al* (1981a). In their case, Ge–S glasses were examined instead of the Ge–Se glasses. Illustration of the OH and SH reduction process is given in figure 4.50. Glasses are prepared using pure Ge (purity: 99.999 999 9%) and pure S (purity: 99.999%). Reagents to yield about 50 g of glass and 5–10 g of supplementary sulfur, to compensate for the loss of evaporation during the heating process, are weighed out into a 3 mm thick silica glass tube. Thus in this method Ge is also heat-treated together with the chalcogen elements. The silica glass tube is then inserted into a furnace and heated to 400–500 °C for 0.5–1 h under an Ar gas flow of 1–2 l min^{-1} or under S_2Cl_2 vapor carried by an Ar gas flow of 1–2 l min^{-1}. Only the sulfur is melted by heat treatment. The melted sulfur in the silica glass tube is cooled to room temperature after heat treatment for 0.5–1 h. Then, the silica glass tube is sealed under a vacuum.

The effect of S_2Cl_2 vapor on the OH and SH reduction is shown in figure 4.51. In the figure, A is the transmission spectrum for the glass

(2.2 mm thick) with no reduction treatment, B for the sample which was heated under an Ar gas flow and C for the sample which was heated under S_2Cl_2 vapor flow. The absorption peaks at 4, 5 and 6.6 μm wavelengths in spectrum A are attributed to SH, GeH, and H_2O impurities, respectively. The heights of these absorption peaks are lowered by heat treatment under an Ar gas flow (spectrum B). Note, however, that a remarkable decrease in the degree of impurity absorption is observed for glasses that have been given the S_2Cl_2 treatment.

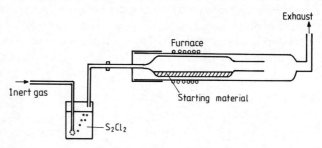

Figure 4.50 Illustration of the apparatus for the reduction of hydrogen impurities in starting materials (Shibata *et al* 1981a).

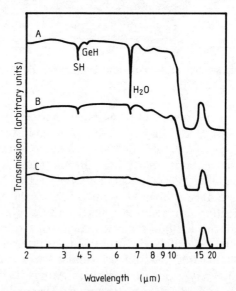

Figure 4.51 Transmission spectra for GeS_3 glasses (2.2 mm thick) which were treated by the following methods: A, no treatment; B, heating under an Ar gas flow; C, heating under S_2Cl_2 vapor flow. The absorption peaks at 4, 5, and 6.6 μm originate from SH, GeH and H_2O impurities, respectively (after Shibata *et al* 1981a).

In addition to the above-mentioned methods, other purification methods have been reported by Hilton *et al* (1975) and Savage *et al* (1978).

Chemical vapor deposition. The fabrication method based on chemical vapor deposition has two main advantages. One is that the homogeneity of the fabricated glass is expected to be high because of the gas-phase mixing of the starting materials. The other is that the possibility of the impurity incorporation during the fabrication process is very low because the glass can be prepared without the exposure of the material to the atmosphere.

First, chemical vapor deposition of Ge–Se binary glass as reported by Katsuyama *et al* (1986) is described here.

In order to fabricate Ge–Se glass by means of chemical vapor deposition, the starting materials must be gases or liquids with high enough vapor pressure. Hydrides and halides of Ge and Se satisfy this requirement, and $GeCl_4$ and $SeCl_2$ are chosen because of their high stabilities. $GeCl_4$ is a liquid at room temperature, and $SeCl_2$ can be obtained by decomposition of Se_2Cl_2, which is also a liquid at room temperature. Furthermore, H_2 is introduced with the reactant gases $GeCl_4$ and $SeCl_2$ in order to improve the reaction efficiency.

In order to avoid contamination by impurities such as oxygen and water, care must be taken to exclude such impurities from the reaction process. Therefore, the reaction and subsequent melting of the glass should be carried out in the same chamber.

The set-up proposed for this purpose is schematically shown in figure 4.52. As shown, the reactant gases $GeCl_4$ and $SeCl_2$, together with H_2, flow into the silica glass tube with Ar used as a transport gas. The reaction is then started by heating the electric furnace. The temperature is set and maintained at 800 °C. The fine Ge–Se particles are then deposited on the inner surface of the silica tube. After deposition, the tube is evacuated at a vacuum of about 10^{-3} Torr and sealed at the neck parts marked A and B in the figure. Next, the sealed

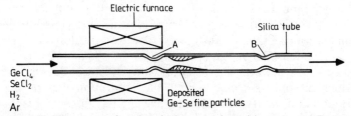

Figure 4.52 The set-up for chemical vapor deposition. A and B are neck parts used for sealing off and separating (after Katsuyama *et al* 1986).

glass tube with the Ge–Se particles is heated at 800 °C for 35 h in another furnace to mix the constituents. The Ge–Se glass block is obtained by air quenching the glass melt in the sealed silica glass tube.

The transmission spectrum of Ge–Se glass with the composition Ge : Se = 20 : 80 (mol.%) is shown in figure 4.53. The absorption band at a wavelength of 12.8 μm is due to the Ge–O lattice vibration. The large absorption band at wavelengths longer than 17 μm is due to intrinsic Ge–Se lattice vibration. It is clear from the figure that a high level of oxygen impurities still remains in the glass. The oxygen content can be estimated to be about 14 wt ppm from the absorption coefficient at 12.8 μm.

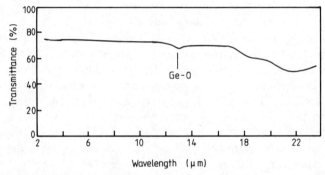

Figure 4.53 The transmission spectrum for Ge–Se glass of composition Ge : Se = 20 : 80 (mol.%). The glass is fabricated by means of chemical vapor deposition, and the sample thickness is 285 μm (after Katsuyama *et al* 1986).

The tail of the 12.8 μm absorption band causes a large increase in transmission loss at 10.6 μm wavelength, which is the most interesting wavelength region for infrared transmission applications. Hence, the oxygen impurities, especially those contained in the Ar transport gas, should be eliminated to achieve low absorption. Furthermore, it is possible to reduce the oxygen impurities in the glass through heat treatments using CO and NH_3 gases. The transmission spectrum of the CO gas treated Ge–Se glass (Ge : Se = 20 : 80 mol.%) is shown in figure 4.54. As shown, the absorption at 12.8 μm due to the Ge–O vibrations disappears completely, while absorption due to the Ge–Se lattice vibration appears at 14 μm wavelength. On the other hand, the transmission spectrum of Ge–Se glass (Ge : Se = 20 : 80 mol.%) treated by NH_3 gas accompanied by subsequent heat treatment is shown in figure 4.55. It can be seen from the figure that the absorption peak at

12.8 μm wavelength nearly disappears, although the absorption peak at 4.6 μm remains to a limited extent. The residual oxygen content in both CO and NH$_3$ treatments is below 1 wt ppm. This is 14 times smaller than that of the untreated glass sample.

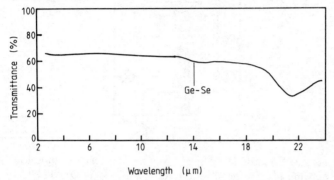

Figure 4.54 The transmission spectrum of CO gas treated Ge–Se glass (Ge : Se = 20 : 80 mol.%). The sample thickness is 653 μm (after Katsuyama *et al* 1986).

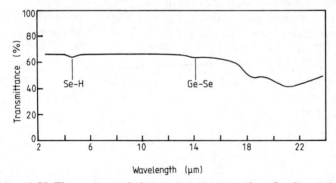

Figure 4.55 The transmission spectrum of Ge–Se glass (Ge : Se = 20 : 80 mol.%), which is treated by NH$_3$ gas and then heated under an Ar gas flow at 500 °C. The sample thickness is 425 μm (after Katsuyama *et al* 1986).

Blanc and Wilson (1985) reported the plasma-enhanced chemical vapor deposition for Ge–Se binary glass fabrication, the apparatus for which is shown in figure 4.56. GeCl$_4$ and Se$_2$Cl$_2$ are used as high purity sources with Ar carrier gas. The plasma is produced in a quartz reaction tube evacuated by a rotary pump via a liquid nitrogen trap. Liquid reactants are contained in temperature-controlled, high purity quartz bubblers. The RF generator is a 2 MHz, 1.5 kW unit.

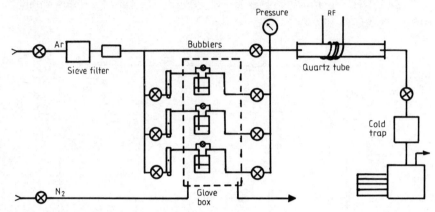

Figure 4.56 A schematic of the liquid sources and reaction tube for RF plasma-enhanced chemical vapor deposition (after Blanc and Wilson 1985).

Infrared transmission in the range 4000–200 cm^{-1} was measured using a thin layer deposited on a crystalline Si substrate (figure 4.57). The largest absorption arises at 800 cm^{-1} due to the Ge–O bond stretching vibration. Further, weaker absorption bands are seen at 781 cm^{-1} (Ge–O). Contributions to the multiple absorption bands towards longer wavelengths come from Ge–Se (256 cm^{-1}), lattice vibrations (741 cm^{-1}), amorphous Ge defects (560 cm^{-1}), Ge–H (570 cm^{-1}) and Ge–Cl (320 cm^{-1}). Although, at the present stage of research, the chemical vapor deposition method does not give excellent transmission properties, the absorption loss due to impurities as well as the scattering loss will be reduced further.

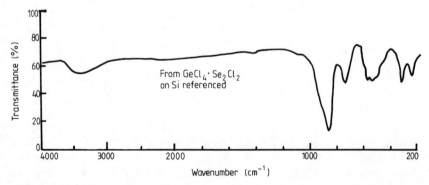

Figure 4.57 Infrared transmission for Ge–Se glass in the range 4000–200 cm^{-1} (after Blanc and Wilson 1985).

4.3.3.2 Fiber drawing

Three fiber drawing methods have been reported so far: the preform method (which is also called rod drawing), the crucible method and the nozzle method.

The preform method. Kapany and Simms (1965) reported the preform method for the first time. Recently, Shibata *et al* (1981b) have presented an advanced version. Sulfide glass optical fibers 150–200 μm in diameter were fabricated from glass rods prepared in the sealed silica ampoules. Figure 4.58 shows the drawing apparatus. The fibers are drawn at a speed of 5–10 m min^{-1}. The temperature in the furnace for fiber drawing must be precisely controlled, because above a certain temperature the glass sublimes and evolves sulfur vapor. Thus the glass fibers are drawn using a tube furnace with a narrow heat region under an Ar gas flow of 1–2 l min^{-1}. The maximum temperature in the heat zone is 300–500 °C and the temperature gradient around the maximum temperature is about 50 °C cm^{-1}.

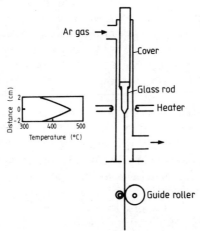

Figure 4.58 Apparatus for fiber drawing and typical data for the temperature gradient at the neck region of the glass rod (after Shibata *et al* 1981b).

Two different coating techniques have been examined. In the first, silicone resin is coated onto the fiber surface through a cylindrical die after drawing and is then cured in a drying furnace at about 500 °C. In the second, glass rods which are inserted into a Teflon FEP tube (14 mm inner diameter, 16 mm outer diameter) are drawn into fibers.

The diameter of the glass fibers is about 150 μm and the thickness of plastic coating layer is about 100 μm for silicone resin or 15 μm for Teflon FEP.

Appropriate plastics for fiber coating are selected according to the thermal properties of chalcogenide glasses (Shibata *et al* 1981b). The softening point of Teflon FEP is 253–282 °C. Since the glass fiber is drawn at a certain temperature between the deformation temperature and the crystallization temperature, Teflon FEP can be used as coating plastic for GeS$_x$ ($x > 4$) (the deformation temperature of that glass is lower than 300 °C) or germanium sulfide glasses containing phosphorus, such as Ge$_3$S$_9$P. On the other hand, the silicone resin must be cured in the tube furnace at more than 500 °C after passing through a coating die. If GeS$_x$ ($x > 4$) or Ge$_3$S$_9$P glasses are coated with silicone resin, evaporation occurs during the cure process. Silicone resin is suitable for GeS$_x$ ($x < 3$) glasses.

However, the most serious problem in fiber drawing using chalcogenide glasses is the sublimation or evolution of the vapor of the chalcogen elements at a temperature above the softening point. The degree of the evolution depends strongly on glass composition and drawing temperature, and is also influenced by the following factors: rod diameter, rate of gas flow, speed of fiber drawing, heater shape and applied power, etc. The appropriate glass compositions for drawing the low loss Ge–P–S glass fibers are shown in figure 4.43. Some composition glasses outside the shaded region cannot be drawn into low loss fibers. Evolution of glass components from the rod causes surface irregularity and increases the optical loss of the drawn fibers. It can be seen from the figure that the composition area suitable for the fiber drawing is surprisingly limited to a region much narrower than the glass forming region.

Saito *et al* (1985) reported the infrared image guiding properties of bundled As–S glass fibers. The fabrication method is, in principle, based on the preform method, and divided into two stages. First, an As–S glass rod is prepared by melting in a sealed ampoule and inserted into a heat-contractile Teflon FEP tube, which is heated in a vacuum so that it might contact and stick to the rod. The preform is heated to 200–300 °C and drawn into a fiber. Secondly, the drawn fiber is cut into sticks 10–30 cm long. Next, these 200–1000 sticks are bundled in a heat-contractile Teflon FEP tube, as shown in figure 4.59. This tube enclosing the bundled fibers is heated and drawn again to obtain the final product: the infrared image guide. Note that the preform can be drawn homogeneously, since As–S glass and Teflon FEP have almost the same softening temperature. By this procedure infrared image guides as long as 10 m can be fabricated. Both of the guide ends are polished carefully with polishing sheets. The core diameter of each fiber element is 0.09 mm and the bundle diameter is 2.0 mm.

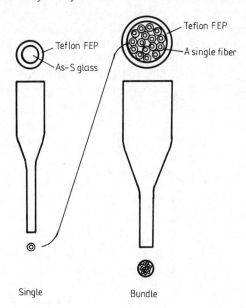

Teflon FEP

Teflon FEP

As-S glass

A single fiber

Single

Bundle

Figure 4.59 The fabrication method for an infrared image guide (Saito *et al* 1985).

The crucible method. The advantage of the crucible method is that there is no strict requirement for the preworking of glass into tube and rod forms. Furthermore, since the glass is heated to a liquid melt form for several hours, there are no annealing requirements.

Kapany and Simms (1965) described the crucible method in detail. Figure 4.60 shows the quartzware concentric assembly used for drawing As–S glass fibers mounted in its electrical furnace. The As–S is thoroughly cleaned and all dust and fractured pieces removed to reduce the density of potential seeding nuclei before it is placed in the crucible. The crucible is loaded with core glass in the upper section and clad glass in the lower section and is then fitted into the furnace and heated to 500 °C for several hours. The inside of the furnace and the crucible are flushed with N_2 gas during this stage to prevent the formation of arsenic and sulfur oxides. During this stage the bubbles escape and the seeds disappear. The temperature of the assembly is then reduced to about 400 °C and left to stabilize for a period of 2 h. Throughout these phases of the operation the drawing nozzle at the bottom of the crucible is kept cold so that a solid slug forms above it and prevents the molten glass in the crucible from escaping. The draw is begun by heating the nozzle under careful control so that the coating glass descends down the outer ring of the nozzle and the core glass down the central tube. The flow of glass must be slow enough initially to keep it under close control, yet the glass must be hot enough to leave the nozzle and form a small drawing cone beneath it. The draw is not considered to be satisfactorily underway until a stable symmetrical cone of flowing glass about 0.75 cm long exists beneath the nozzle. The ratio of core glass to clad glass can be varied by raising or lowering the crucible with respect to the upper one; however, great care is needed to maintain the concentricity of the nozzles.

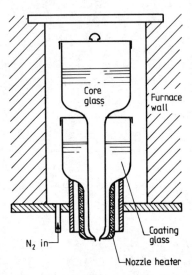

Figure 4.60 The concentric crucible apparatus for drawing single coated As–S glass fibers (after Kapany and Simms 1965).

Recently, Kanamori *et al* (1984) also reported the crucible method, which is described in figure 4.61. The drawn As–S core-clad fibers, with a cladding–core diameter ratio of 1.4–2, are 100–300 μm in outer diameter and 300–1000 m long. However, it should be noted that the fiber drawing region is, in general, narrower than the preform method, as shown in figure 4.42.

The nozzle method. The nozzle method is a modification of the crucible method. The drawing apparatus is shown schematically in figure 4.62 (Katsuyama *et al* 1984). The glass blocks are drawn into fibers from the quartz nozzle under a pressure of 0.5–1.5 atm. This applied pressure drawing process enables us to fabricate fibers under relatively low temperatures, and so the crystallization of the chalcogenide glass during the drawing process can be minimized. This leads to the reduction of scattering loss of the optical fibers.

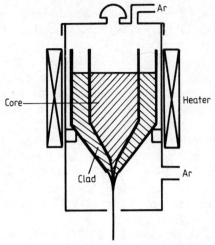

Figure 4.61 Double-crucible assembly for drawing As–S glass core-clad fibers (after Kanamori *et al* 1984, © 1984 IEEE).

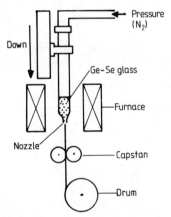

Figure 4.62 The drawing apparatus for the nozzle method (after Katsuyama *et al* 1984).

4.3.4 Properties

4.3.4.1 Loss characteristics

Sulfide glass fibers. Kapany and Simms (1965) obtained spectral transmission data on As–S fibers for the first time, which showed, however, relatively high loss. Recently, several reports on As–S, Ge–S and Ge–P–S glass fibers have been published. These sulfide glass fibers were studied mainly for long haul optical transmission lines, because of the possibility of low loss at 2–4 μm wavelength. Furthermore, these fibers are used for the transmission line of CO laser light power (wavelength 5.3 μm).

Kanamori *et al* (1984, 1985) have studied As–S glass fibers extensively. Measured transmission losses versus photon energy for the $As_{40}S_{60}$ bulk glass (short wavelength region) and optical fiber (long wavelength region) are shown in figure 4.63. In the short wavelength region, the loss consists of two exponential parts. The steeper region corresponds to the Urbach tail, which originates from the electronic transition between the valence band and conduction band. The less steep region is due to the weak absorption tail (Wood and Tauc 1972). The transmission loss in the wavelength region longer than 6 μm is mainly due to multiphonon absorption of the glass matrix. Absorption bands at 14.5, 12.1, 10.2, 9.5, 7.6 and 6.9 μm can be explained by the intrinsic vibration modes in the glass matrix. The absorption peaks in the 1.4–6.4 μm region are caused by several impurities. The strong absorption bands at 2.91 and 4.03 μm are due to the OH and SH fundamental stretching vibrations, respectively. Absorption bands at 6.32 and 2.77 μm are due to molecular H_2O. Bands at 4.94 and 5.13 μm are associated with carbon and oxygen, respectively.

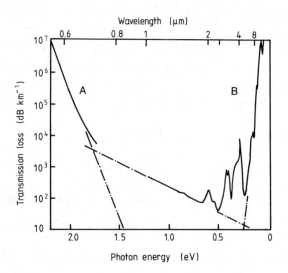

Figure 4.63 Transmission loss versus photon energy for $As_{40}S_{60}$ bulk glass (A) and unclad fiber (B) (after Kanamori *et al* 1984, © 1984 IEEE).

Of these losses, the weak absorption loss is the most important for predicting the minimum transmission loss attainable for chalcogenide glass fibers (Kanamori *et al* 1985). It is thought that the weak absorption is induced by additional band gap states and the magnitude of the tail depends on the total concentration of states. Additional band gap states which cause a weak absorption tail are, in general, due to impurities such as transition metal elements or defects such as valence alternation pairs and dangling bonds. In homogeneous glass with high purity, the gap states are considered to be due to the defects formed at the glass transition temperature of the glass in the melt cooling process. Weak absorption tails are observed in chalcogenide glass fibers with high purity but not in oxide glass fibers. This indicates that the defect concentration in chalcogenide glass is very high and the formation energy of a defect in the glass is fairly small compared to those for oxide glass. Therefore, since the weak absorption tails observed in the chalcogenide glass fibers are related to the intrinsically small formation energy of a defect in the glasses, it is difficult to reduce the magnitude of the weak absorption tails much below those indicated in figure 4.63, and it is estimated that weak absorption tails inhibit transmission loss reduction to the level of below $10 \, \text{dB km}^{-1}$ at the lowest loss wavelengths in chalcogenide glass fibers. Therefore, Kanamori *et al* (1985) concluded that these sulfide glass optical fibers are not appropriate for use in long haul optical communication systems.

As–S glass optical fibers have also been studied by Vechkanov *et al* (1982), Miyashita and Terunuma (1982), Vasiliev *et al* (1983), Hattori *et al* (1984), Andriesh *et al* (1984) and Arai and Kikuchi (1984).

Ge–S and Ge–P–S glass systems for infrared fibers have been studied by Shibata *et al* (1980b, 1981a,b) and Kanamori *et al* (1984). Figure 4.64 shows transmission loss versus photon energy for bulk glass (short wavelength region) and unclad fibers (long wavelength region) with $Ge_{20}S_{80}$ (Kanamori *et al* 1984). In the figure the solid line shows the loss curve for the fiber with a lower hydrogen impurity content (I), while the broken line shows the loss curve of the fiber with higher hydrogen content (II). In both spectra, peaks at 2.85 and 4.00 μm originate from OH and SH fundamental stretching vibrations, respectively. The loss curve for the $Ge_{20}S_{80}$(I) fiber shows three optical windows with minimum losses of $500 \, \text{dB km}^{-1}$ at 2.40 μm, $560 \, \text{dB km}^{-1}$ at 3.33 μm, and $840 \, \text{dB km}^{-1}$ at 4.60 μm. In the loss curve for the $Ge_{20}S_{80}$(II) fiber, the lowest loss of $148 \, \text{dB km}^{-1}$ is observed at 1.68 μm.

A large difference exists in the loss spectra for the two fibers. The loss for the (II) fiber is lower in the short wavelength region, but is higher in the long wavelength region than for the (I) fiber. The higher loss in the long wavelength region for $Ge_{20}S_{80}$(II) is clearly due to the high level of hydrogen impurity. On the other hand, the short wavelength behavior is attributed to the weak absorption tail, which originates from the gap states caused by dangling bonds. The difference in magnitude of the weak absorption tails shown in the figure indicates that hydrogen impurities may decrease the number of gap states.

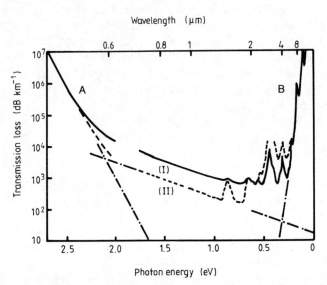

Figure 4.64 Transmission loss versus photon energy for $Ge_{20}S_{80}$ bulk glasses (A) and unclad fibers (B). The solid line shows the loss curve for a fiber prepared from a glass with a low hydrogen impurity content (I), while the broken line shows the loss curve of a fiber with a high hydrogen content (II) (after Kanamori *et al* 1984, © 1984 IEEE).

Shibata *et al* (1980b) studied $Ge_3PS_{7.5}$ glass fibers, whose loss spectrum is shown in figure 4.65. The glass fiber has an optical loss of $0.4 \, dB \, m^{-1}$ at about $2.5 \, \mu m$. The G–P–S glass is thought to have a relatively stable vitreous state.

An optical fiber with an As_2Se_3 core and As_2S_3 cladding has been fabricated by Vasiliev *et al* (1983). The optical loss is about $1 \, dB \, m^{-1}$ at $3.5 \, \mu m$; the value is greater than for the unclad fiber, and is probably due to either the insufficient purity of the inert atmosphere during fiber drawing or the insufficient accuracy of the double-layer fiber drawing process.

Selenide glass fibers. Selenide glass fibers can be classified into As–Se based glass fibers and Ge–Se based glass fibers. Binary As–Se glass fibers have been studied by Vlasov *et al* (1982), Vechkanov *et al* (1982), Vasiliev *et al* (1983) and Bornstein *et al* (1985). Transmission loss of less

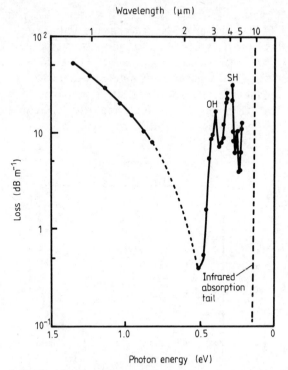

Figure 4.65 The loss spectrum for $Ge_3PS_{7.5}$ unclad glass fiber, of diameter 160 μm and length 0.5 m. The dashed line at 1.5–2.4 μm wavelengths shows the extrapolation of the loss curve at 0.8–1.5 μm wavelengths. The dashed line at about 10 μm shows the infrared absorption tail which is obtained by the extrapolation of measured loss values for glass plates (2–5 mm thick) (after Shibata *et al* 1980b).

than 1 dB m^{-1} at wavelengths of 3–5 μm was obtained by Vasiliev *et al* (1983).

Ternary As–Ge–Se glass fibers have recently been studied by Takahashi *et al* (1983) and Kanamori *et al* (1984, 1985). Figure 4.66 shows the transmission loss spectrum obtained for $As_{38}Ge_5Se_{57}$ bulk glass (< 0.7 μm) and optical fiber (> 1 μm). The loss spectrum consists of the Urbach tail (< 0.7 μm), weak absorption tail (0.7–2 μm), impurity absorption peaks and multiphonon absorption (> 2 μm). A band at 20.6 μm is attributed to the two-phonon process of the As–Se fundamental stretching vibration (Maklad *et al* 1974). Strong absorption bands at 2.92 and 4.57 μm are also identified as due to the OH and SeH fundamental stretching vibrations, respectively (Moynihan *et al* 1975). Absorption bands at 7.90, 9.52, 12.8 and 14.9 μm are due to oxygen impurities. Absorption bands at 6.32, 2.83 and 2.76 μm are due to molecular H_2O and that at 4.94 μm to carbon impurity. Kanamori *et al* thought that the band at 7.05 μm was probably associated with impurities, although an exact identification cannot be made. Impurity absorption bands of As–Ge–Se glasses assigned by Kanamori *et al* (1984) are summarized in table 4.15. Impurity absorption bands of the As–S and Ge–S glasses are also summarized for comparison. The minimum loss of the $As_{38}Ge_5Se_{57}$ glass fiber is 182 dB km^{-1} at 2.12 μm.

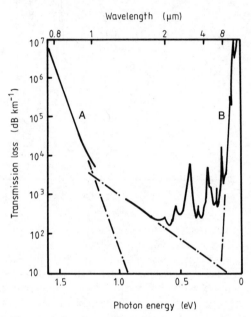

Figure 4.66 Transmission loss versus photon energy for $As_{38}Ge_5Se_{57}$ bulk glass (A) and unclad fiber (B) (after Kanamori *et al* 1984, © 1984 IEEE).

The most interesting application of selenide glass fiber is in CO_2 laser light transmission, of wavelength 10.6 μm. Figure 4.67 shows the transmission loss (10.6 μm wavelength) versus temperature obtained by Takahashi *et al* (1983). The transmission losses at 100 °C, room and liquid N_2 temperatures are 10.9, 4.5 and 0.5 dB m^{-1}, respectively. The loss depends strongly on temperature and increases with it. It is known that the temperature change causes the shift of the multiphonon absorption band and the loss at the tail of the absorption decreases with decreasing temperature. The experimental result is due to the fact that 10.6 μm wavelength meets just at the edge of the intrinsic multiphonon

Table 4.15 Impurity absorption bands of chalcogenide glasses (after Kanamori *et al* 1984, © 1984 IEEE).

Cause		Wavelength (μm)		
		As–S	As–Ge–Se	Ge–S
OH	Fundamental	2.91	2.92	2.84
	Overtone	1.44	1.45	1.43
	Combination	2.29, 1.92	2.32, 1.92	2.25, 1.88
SH	Fundamental	4.03	4.57	4.00
(SeH)	Overtone	2.05	2.32	2.02
	Combination	3.69, 3.11, 2.54	4.15, 3.55	3.78, 3.08
H_2O		6.32, 2.77	6.32, 2.76	6.32, 2.76
C		4.94	4.94	4.94
Oxide		–	7.90	–

absorption band. It was found from the experiment that the temperature change of transmission loss for a selenide fiber is larger than that for other materials. In general, the bond energy in chalcogenide glass is small. This would be the reason why the selenide glass shows strong temperature dependence of the transmission loss.

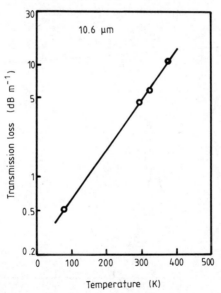

Figure 4.67 Transmission loss at 10.6 μm versus temperature for As–Ge–Se glass fiber (after Takahashi *et al* 1983).

Takahashi *et al* (1983) also studied CO_2 laser power transmission. Transmitted CO_2 laser power through a 1 m long fiber was 1.2 W at the temperature of liquid N_2.

Binary Ge–Se glass optical fibers have been studied by Katsuyama *et al* (1984). The loss spectrum of a $Ge_{20}Se_{80}$ glass fiber is shown in figure 4.68. Results show that the transmission loss is less than $1\,dB\,m^{-1}$ for the wide infrared region. Absorption due to SeH occurs to a small extent. The minimum loss is $0.2\,dB\,m^{-1}$ at $5.5\,\mu m$. Figure 4.68 also shows that the loss is relatively high in the wavelength region above $9\,\mu m$. The loss at $10.6\,\mu m$ (CO_2 laser wavelength) is as high as $8\,dB\,m^{-1}$. These losses are mainly due to intrinsic Ge–Se lattice vibrations. Katsuyama *et al* (1984) found that the addition of Sb is effective in lowering lattice vibration. Figure 4.69 shows the absorption coefficients of various Ge–Sb–Se bulk glasses for wavelengths longer than $9\,\mu m$. Note that the absorption band around $13\,\mu m$ strongly affects the transmission loss at $10.6\,\mu m$ and the peak height decreases as the Sb content increases. Therefore, it can be expected that the addition of Sb eliminates the influence of the lattice vibration on the $10.6\,\mu m$ loss. $Ge_{10}Sb_{26}Se_{64}$ glass, which shows a relatively small absorption peak at $13\,\mu m$, has been drawn into a fiber. Figure 4.70 shows its transmission loss spectrum in the wavelength region 5–$25\,\mu m$. The result of the $Ge_{20}Se_{80}$ glass fiber is also shown for comparison. The result shows that the loss at $10.6\,\mu m$ decreases to $3\,dB\,m^{-1}$. Furthermore, the wavelength region for the loss of less than $1\,dB\,m^{-1}$ reaches up to $10\,\mu m$.

Ge–Se based glass optical fibers have been studied by Boniort *et al* (1980), Brehm *et al* (1982), Le Sergent (1982) and Dianov *et al* (1984). The experimental results are summarized in Table 4.14.

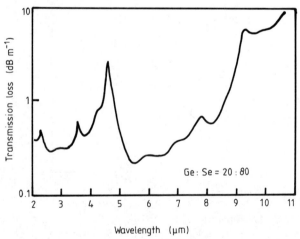

Figure 4.68 The transmission loss spectrum for $Ge_{20}Se_{80}$ chalcogenide glass fiber (after Katsuyama *et al* 1984).

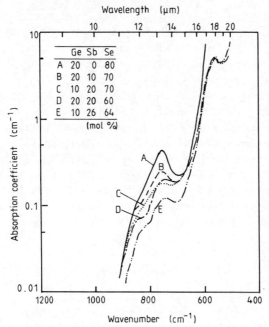

Figure 4.69 Absorption coefficients of various Ge–Sb–Se bulk glasses (after Katsuyama *et al* 1984).

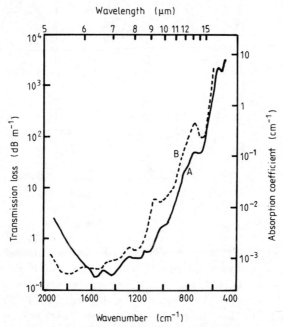

Figure 4.70 Transmission loss spectra for $Ge_{10}Sb_{26}Se_{64}$ (A) and $Ge_{20}Se_{80}$ (B) glass optical fibers (after Katsuyama *et al* 1984).

Telluride glass fibers. Telluride glass fibers are thought to be advantageous in lowering the loss at a 10.6 μm wavelength because the heavy Te atoms have the effect of shifting the absorption band due to lattice vibration toward longer wavelengths. Figure 4.71 shows the absorption coefficients for wavelengths above 10 μm, which were measured by Katsuyama and Matsumura (1986). These absorptions are caused by an intrinsic lattice vibration. Results indicate that as the Te content increases the infrared absorption edge shifts towards longer wavelengths. Moreover, the absorption peak at 13 μm decreases. The transmission loss is therefore expected to be less than 1 dB m^{-1} at 10.6 μm for a Te content of more than 50 mol.% (curves C and D in figure 4.71). Curve D in particular indicates the possibility of an ultra-low loss of less than 0.1 dB m^{-1}. However, as the Te content increases, it may become more difficult to draw into glass fibers without yielding microcrystals.

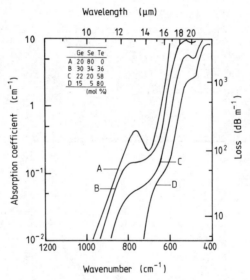

Figure 4.71 Absorption due to lattice vibration for various Ge–Se–Te ternary glass compositions (after Katsuyama and Matsumura 1986).

To avoid crystallization, a high speed drawing method has been developed. Glass blocks are drawn into fibers from the quartz nozzle under N$_2$ pressure of around 1 atm. The drawing speed is controlled to be as high as 1–5 m s^{-1}. The transmission loss of the Ge$_{22}$Se$_{20}$Te$_{58}$ glass fiber fabricated by this method is shown in figure 4.72. The loss at wavelengths below $2\,\mu$m is due to the Urbach tail, while that at wavelengths above $10\,\mu$m is due to lattice vibration. The small loss at $9.3\,\mu$m is thought to be caused by Si impurities. In contrast, the broad band loss in the wavelengths between 2 and $10\,\mu$m is attributed to Mie scattering caused by the few microcrystals which remain in the glass. Mie scattering is defined as scattering caused by scattering centers whose dimensions are of the same order of magnitude as the light wavelength. The wavelength dependence of this loss is shown in figure 4.73. The figure confirms that loss changes linearly with the wavelength to the power -3. It is also found that the loss increases as the Te content increases. The microcrystals causing the Mie scattering are thought to be hexagonal Te. The transmission loss at $10.6\,\mu$m can be reduced to as low as $1.5\,\mathrm{dB\,m^{-1}}$, as shown in figure 4.72.

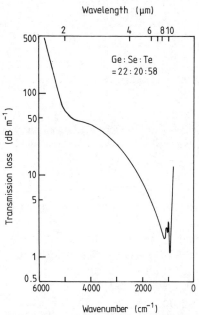

Figure 4.72 The transmission loss spectrum for $\mathrm{Ge_{22}Se_{20}Te_{58}}$ glass optical fibers (after Katsuyama and Matsumura 1986).

4.3.4.2 *Refractive index and dispersion*

Refractive indices of chalcogenide glasses are, in general, larger than those of oxide and halide glasses. Refractive indices of the binary $\mathrm{As_2S_3}$ and $\mathrm{GeS_3}$ glasses are 2.41 (at $5.3\,\mu$m (Rodney *et al* 1958)) and 2.113 (n_D (Shibata *et al* 1981b)), respectively. Ternary As–S based and Ge–S based glasses have almost the same refractive indices as those of the binary glasses.

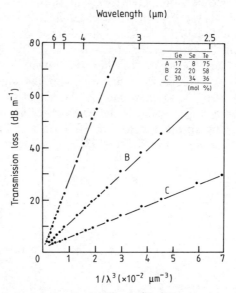

Figure 4.73 Transmission loss versus $1/\lambda^3$ (λ is wavelength) for three Ge–Se–Te glass optical fibers (after Katsuyama and Matsumura 1986).

Typical binary selenide glasses have refractive indices of 2.411 ($Ge_{20}Se_{80}$, at 5.4 μm (Aio *et al* 1978)) and 2.7840 (As_2S_3, at 4.0 μm (Webber 1976)). Some ternary Se-based chalcogenide glasses have refractive indices of 2.62 ($Ge_{28}Sb_{12}Se_{60}$, at 5 μm (Hilton 1973)) and 2.49 ($Ge_{33}As_{12}Se_{55}$, at 5 μm (Hilton 1973)).

There are only a few reports on material dispersion. Material dispersion of As_2S_3 glass optical fibers calculated from the data reported by Rodney *et al* (1958) is shown in figure 4.74 (Miyashita and Manabe 1982). Material dispersion of As_2S_3 glass fibers falls to zero at 4.89 μm, which fortunately coincides with the loss minimum wavelength. It is also noteworthy that the slope of the curve is relatively lower than those for oxide or fluoride glasses.

4.3.4.3 Mechanical and thermal properties

In general, chalcogenide glasses are relatively weak in their mechanical hardness. In addition, glass transition temperatures and crystallization

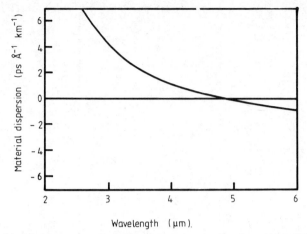

Figure 4.74 Material dispersion curve for As_2S_3 glass (after Miyashita and Manabe 1982, © 1982 IEEE).

temperatures are lower than those of oxide glasses and fluoride glasses. Therefore, much care is needed in the handling of chalcogenide glass optical fibers. For example, in the case of CO_2 laser energy transmission, the temperature increase due to transmitted high power light should be prevented by water cooling.

Densities, glass transition temperatures, crystallization temperatures, thermal expansion coefficients and hardnesses of the typical chalcogenide glasses are summarized in tables 4.16 and 4.17. In the tables, glass compositions were chosen which were actually used for the infrared optical fibers. Tl 1173 and Tl 20 in table 4.17 are the glass compositions which were developed by Hilton *et al* (1975) for use with high energy CO_2 lasers.

4.3.5 Summary

Chalcogenide glass optical fibers studied so far can be divided into two groups: sulfide glass fibers and selenide and telluride glass fibers. The transmission loss of sulfide glass fiber has been reduced to less than $0.1 \, dB \, m^{-1}$ at wavelengths of 2–5 μm (Kanamori *et al* 1985). However, further loss reduction is thought to be difficult because of the existence of the intrinsic weak absorption tail (Kanamori *et al* 1985). Sulfide glass fibers are now used for the short distance transmission of light. CO laser light (wavelength 5.3 μm) can be transmitted through As–S glass fibers. Infrared image transmission is also possible using bundle fibers.

Selenide and telluride glass fibers have mainly been studied for the transmission of the CO_2 laser power at a wavelength of 10.6 μm.

Table 4.16 Physical and chemical properties of sulfide glasses.

Material	Density	Glass transition temperature (°C)	Crystallization temperature (°C)	Softening point (°C)	Thermal expansion coefficient ($\times 10^6$)	Knoop hardness (kg mm^{-2})	References
As$_2$S$_3$	3.2	–	–	205	25	109	Hattori et al (1984), Kruse et al (1962)
GeS$_3$	2.5	260	500	370	25	130	Shibata et al (1981b)
GeS	–	–	–	420	14	179	Le Sergent (1982)

Table 4.17 Physical and chemical properties of selenide glasses.

Material	Density	Glass transition temperature (°C)	Crystallization temperature (°C)	Thermal expansion coefficient ($\times 10^6$)	Knoop hardness (kg mm^{-2})	References
As_2Se_3	4.62	178	286	21	156.2	Webber (1976)
$Ge_{20}Se_{80}$	4.37	154	–	24.8	147.3	Webber (1976)
$Ge_{10}Sb_{25}Se_{65}$	5.04	192	312	18.4	–	Savage et al (1978)
$Ge_5As_{40}Se_{55}$	4.53	196	–	–	–	Webber (1976)
$Ge_{28}Sb_{12}Se_{60}$ (TI 1173)	4.68	277	522	13.79	206	Savage et al (1978)
$Ge_{33}As_{12}Se_{55}$ (TI 20)	4.41	368	539	–	252.4	Webber (1976)
$Ge_{20}Se_{20}Te_{60}$	–	169	260	–	–	Sarrach and De Neufville (1976)

Although transmission losses of about 1 dB m^{-1} were achieved (Katsuyama and Matsumura 1986), further loss reduction is still needed for stable power transmission. In contrast, Ge–Se and As–Se glass fibers have wide infrared windows, and so can be used for transmission lines in infrared spectroscopy.

4.4 Other glass fibers

4.4.1 Introduction

ZnCl$_2$ optical fibers have potential for ultra-low loss at wavelengths above 2 μm (Van Uitert and Wemple 1978). However, ZnCl$_2$ suffers the disadvantage of hygroscopic behavior, which makes fiber fabrication difficult. Therefore at the present stage of research, ZnCl$_2$ glass is not being studied extensively. Glass materials such as ZnBr$_2$, AgI–AgF–AlF$_3$, AgBr–CsI–PbBr$_2$, PbF$_2$–BaF$_2$–InF$_3$, and PbBr$_2$–PbI$_2$ have also been studied.

4.4.2 Materials
4.4.2.1 ZnCl$_2$
ZnCl$_2$ glass has the potential for ultra-low loss in the 3.5–4 μm wavelength region, which was predicted numerically from scattering and lattice absorption losses (Van Uitert and Wemple 1978).

The scattering loss is described by

$$\alpha_s \simeq \lambda^{-4} B^2 Z_a^2 \beta T_g / E_0^4, \tag{4.3}$$

where λ is wavelength, B is a dimensionless structure factor given by $B = N_A d^3$ (N_A is the volume density of anions and d is the first-neighbor bond length), Z_a is the formal chemical valence of the anion, β is the isothermal compressibility, T_g is the glass transition temperature, and E_0 is the average energy gap of the glass. Equation (4.3) reveals that halides whose Z_a is equal to 1 provide an intrinsic factor-of-4 improvement over oxides ($Z_a = 2$). Materials with cations located in tetrahedral sites within a close-packed anion lattice (e.g. ZnCl$_2$) have a small value of B ($= 3\sqrt{3}/16 = 0.325$) compared to materials with cations in octahedral sites ($B = 0.500$) such as NaCl or CdCl$_2$. These features are advantageous for ZnCl$_2$ glass. The equation also shows that a large energy gap is essential for small scattering loss.

Figure 4.75 shows the estimated scattering loss of the ZnCl$_2$ glass together with the lattice vibration edges of the various materials. As shown, ZnCl$_2$ glass is predicted to have a minimum loss value of 0.001 dB km^{-1} in the 3.5–4 μm region. In addition, the calculation of material dispersion shows that it falls to zero in the 3.3–3.7 μm region. This wavelength region is nearly equal to that for minimum loss. On the basis of the above discussion, Van Uitert and Wemple (1978) predicted that $ZnCl_2$ glass can be used for the ultra-low loss optical fibers whose transmission loss exceeds that of silica glass fiber.

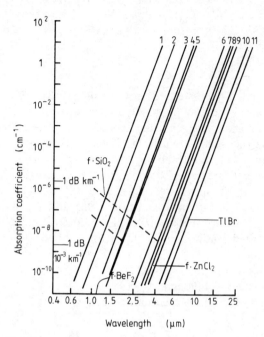

Figure 4.75 Extrapolation of the intrinsic absorption (α) versus λ for (1) fused SiO_2, (2) Al_2O_3, (3) LiF, (4) MgF_2, (5) fused BeF_2, (6) KF, (7) NaCl, (8) $ZnCl_2$, (9) KCl, (10) KBr, and (11) TlBr. The arrows indicate the wavelengths giving zero material dispersion for various materials. The upper dashed line indicates known scatter loss values for fused SiO_2 as well as the predicted values for $ZnCl_2$. The lower dashed curve represents the predicted values for BeF_2 (after Van Uitert and Wemple 1978).

Following the prediction by Van Uitert and Wemple, several investigations into the fabrication of $ZnCl_2$ glass have been carried out. Robinson *et al* (1982) prepared pure and doped $ZnCl_2$ glasses using a reactive atmosphere process (RAP). CCl_4 was used as the RAP agent, resulting in the reduction of H_2O impurities. However, the introduction of the dopants such as $GaCl_3$ was found to be ineffective in stabilizing the glass state. Robinson *et al* (1982) suggested that $ZnCl_2$ glass is basically applicable to low loss transmission at a CO_2 laser wavelength. However, further improvements in the hygroscopic nature of the glass are required.

Recently, Yamane *et al* (1985) prepared a new $ZnCl_2$-based glass of composition $ZnCl_2(48)$–$KBr(48)$–$PbBr_2(4)$. The glass transition temperature T_g, crystallization temperature T_x, softening temperature T_s and linear thermal expansion coefficient of the glass are 45–46 °C, 100 °C, 54 °C and 570×10^{-7} °C^{-1}, respectively. The refractive index at 632.8 nm is 1.63. The transmission loss at 10.6 μm, however, remained as large as 20 dB m^{-1}.

4.4.2.2 Other glasses

$ZnBr_2$ glasses have been studied by Hu *et al* (1983). They were found to have a transparent region up to 20 μm in wavelength, which is much longer than fluoride and chloride glasses. However, bromide glasses show the same hygroscopic nature as $ZnCl_2$ glasses, and so this problem must be solved if they are to be used for optical fibers.

Komatsu *et al* (1985) fabricated AgI–AgF glasses containing a small amount of AlF_3. These glasses showed good infrared transparency to about 10 μm except for an absorption band at 7.6 μm. They dissolve slowly in water but are less hygroscopic than $ZnCl_2$ and $ZnBr_2$ glasses. However, detailed studies on the fabrication of the fiber have not yet been carried out.

Glass formation in the $BiCl_3$–KCl system (Angell and Ziegler 1981), $PbBr_2$–PbI_2 system (Hidaka *et al* 1980), AgI–CsI–$PbBr_2$ (Nishii *et al* 1985) and PbF_2–BaF_2–InF_3 (Auriault *et al* 1985) has also been reported.

4.4.3 Summary

Some halide glasses, such as $ZnCl_2$ glasses, are potentially useful materials both for ultra-low loss fibers in the 3.5–4 μm wavelength region and for fibers transmitting the CO_2 laser light at 10.6 μm wavelength. However, their glass properties are inferior to fluoride and chalcogenide glasses. In particular the problem of the hygroscopic nature of these glasses must be solved before they can be used as fiber materials.

References

Aio L G, Efimove A M and Kokorina V F 1978 *J. Non-Cryst. Solids* **27** 299–307
Andriesh A M, Bol'shakov O V, Lungu D N, Ponomev V V, Smirnova A S and Fedotova N D 1984 *Sov. J. Quantum Electron.* **14** 855–7
Angell C A and Ziegler D C 1981 *Mater. Res. Bull.* **16** 279–83
Arai T and Kikuchi M 1984 *Appl. Opt.* **23** 3017–19
Auriault N, Guery J, Mercier A M, Jocoboni C and De Pape R 1985 *Mater. Res. Bull.* **20** 309–14
Bendow B 1984 *NOSC Final Rep.* Feb. 28

Bendow B, Brown R N, Drexhage M G, Lorentz T J and Kirk R L 1981b *Appl. Opt.* **20** 3688–90

Bendow B, Mitra S and Moynihan C T 1981a *Appl. Opt.* **20** 2875–7

Blanc D and Wilson J I B 1985 *J. Non-Cryst. Solids* **77 & 78** 1129–32

Boniort J Y, Brehm C, Dopont P H, Guignot D and Le Sergent C 1980 *Proc. 6th European Conf. Opt. Commun.* 61–4

Bornstein A, Croitoru N and Marom E 1985 *J. Non-Cryst. Solids* **74** 57–65

Brehm C, Cornebois M, Le Sergent C and Parant J P 1982 *J. Non-Cryst. Solids* **47** 251–4

Brown R and Hutta J J 1985 *Appl. Opt.* **24** 4500–3

Chen D, Skogman R, Bernal E G and Butter C 1979 *Fiber Optics* (New York: Plenum) pp 119–22

Dianov E M 1982 *Advances in IR Fibers, Tech. Dig.* (Los Angeles, CA: SPIE) paper 320–04

Dianov E M, Masychev V J, Plotnichenko V G, Sysoev V K, Baikalov P J, Devjatykh G G, Konov A S, Schipachev J V and Churbanov M F 1984 *Electron. Lett.* **20** 129–30

Drexhage M G, Bendow B, Brown R N, Banerjee P K, Lipson H G, Fonteneau G, Lucas J and Moynihan C T 1982b *Appl. Opt.* **21** 971–2

Drexhage M G, Bendow B, Lipson H and Moynihan C T 1980 *Proc. Boulder Laser Damage Symposium (Boulder, CO)*

Drexhage M G, Bendow B and Lorentz T J 1981 *Tech. Digest 3rd Int. Conf. on Integrated Optics and Optical Fiber Communications* p 32

Drexhage M G, El-Bayoumi O H and Moynihan C T 1982a *Advances in IR Fibers, Tech. Dig.* (Los Angeles CA: SPIE) paper 320–06

Fonteneau G, Lahaie F and Lucas J 1980 *Mater. Res. Bull.* **15** 1143–7

Fonteneau G, Tregoat D and Lucas J 1985 *Mater. Res. Bull.* **20** 1047–51

Ford E B and Savage J A 1976 *J. Phys. E: Sci. Instrum.* **9** 622–4

France P W, Carter S F, Moore M W and Williams J R 1985 *Electron. Lett.* **21** 602–3

France P W, Carter S F, Williams J R and Beales K J 1984 *Electron. Lett.* **20** 607–8

Galeener F L, Mikkelsen J C Jr, Geils R H and Mosby W J 1978 *Appl. Phys. Lett.* **32** 34–6

Gannon J R 1981 *Tech. Digest Soc. Photo-Optical Instrum., Los Angeles, CA* 62–8

Gannon J and Byron K C 1982 *Proc. SPIE* **320** 34

Ginther R J and Tran D C 1981 *Tech. Digest 3rd Int. Conf. on Integrated Optics and Optical Fiber Communications* p 32

Harrington J A, Braunstein M, Bobbs B and Braunstein R 1981 *Physics of Fiber Optics* ed B Bendow and S S Mitra (American Ceramics Society) pp 94–103

Hattori T, Sato S, Fujioka T, Takahashi S and Kanamori T 1984 *Electron. Lett.* **20** 811–2

Hidaka T, Morikawa T and Shimada J 1980 *Tech. Dig. Annual Meeting of the Institute of Electronics and Communication Engineers of Japan* pp 4–186 (in Japanese)

Hilton A R 1973 *J. Electron. Mater.* **2** 211–25

Hilton A R, Hayes D J and Rechtin M D 1975 *J. Non-Cryst. Solids* **17** 319–38

Hilton A R, Jones C E and Brau M 1966 *Phys. Chem. Glasses* **7** 105–12

Hu H, Ma F and Mackenzie J D 1983 *J. Non-Cryst. Solids* **55** 169–72

Izawa T 1977 *Technical Digest of the Conference of Integrated Optics and Optical Fiber Communications (Tokyo)* C1–1 p 373

Kanamori T and Sakaguchi S 1986 *Japan. J. Appl. Phys.* **25** L468–70

Kanamori T, Shibata S, Mitachi S and Manabe T 1980 *Japan. J. Appl. Phys.* **19** L90–2

Kanamori T, Terunuma Y, Takahashi S and Miyashita T 1984 *J. Lightwave Technol.* **LT-2** 607–13

—— 1985 *J. Non-Cryst. Solids* **69** 231–42

Kapany N S and Simms R J 1965 *Infrared Phys.* **5** 69–80

Katsuyama T, Ishida K, Satoh S and Matsumura H 1984 *Appl. Phys. Lett.* **45** 925–7

Katsuyama T and Matsumura H 1986 *Appl. Phys. Lett.* **49** 22–3

Katsuyama T, Satoh S and Matsumura H 1986 *J. Appl. Phys.* **59** 1446–9

Komatsu T, Ur H and Doremus R H 1985 *J. Non-Cryst. Solids* **69** 309–15

Kruse P W, McGlauchlin L D and McQuistan R B 1962 *Elements of Infrared Technology* (New York: John Wiley) pp 140–1

Lau J, Nakata A M and Mackenzie J D 1985 *J. Non-Cryst. Solids* **70** 233–42

Le Sergent 1982 *Tech. Dig. Soc. Photo-Optical Instrum., Los Angeles, CA* paper 320–21

Lu G, Levin K H, Burk M J and Tran D C 1986 *Electron. Lett.* **22** 602–3

Lucas J 1982 *Advances in IR Fibers, Tech. Dig. SPIE*, Jan. paper 320–05

Lucas J, Chantanashinh M, Poulain M and Poulain M 1978 *J. Non-Cryst. Solids* **27** 273–83

Maklad M S, Mohr R K, Howard R E, Macedo P B and Moynihan C T 1974 *Solid State Commun.* **15** 855–8

Matecki M, Poulain M and Poulain M 1983 *Tech. Dig. 2nd Int. Symp. on Halide Glasses (Troy, NY)* paper 27

Maze G, Cardingand V and Poulain M 1984 *J. Lightwave Technol.* **LT-2** 596–9

Mimura Y, Tokiwa H and Shinbori O 1984 *Electron. Lett.* **20** 100–1

Mitachi S and Manabe T 1980 *Japan. J. Appl. Phys.* **19** L313–4

Mitachi S and Miyashita T 1982 *Electron. Lett.* **18** 170–1

Mitachi S, Miyashita T and Kanamori T 1981a *Electron. Lett.* **17** 672–3

—— 1981b *Electron. Lett.* **17** 591–2

Mitachi S, Miyashita T and Manabe T 1981c *Electron. Lett.* **17** 128–9

—— 1982 *Phys. Chem. Glasses* **23** 196–201

Mitachi S, Ohishi Y and Miyashita T 1983a *J. Lightwave Technol.* **LT-1** 67–70

Mitachi S, Ohishi Y, Terunuma Y and Takahashi S 1983b *Research Report of NTT* **32** 2723–36 (in Japanese)

Mitachi S, Sakaguchi S, Yonezawa H, Shikano K, Shigematsu T and Takahashi S 1985 *Japan. J. Appl. Phys.* **24** L827–8

Mitachi S, Terunuma Y, Ohishi Y and Takahashi S 1983c *Japan. J. Appl. Phys.* **22** L537–8

—— 1984 *J. Lightwave Technol.* **LT-2** 587–92

Miyashita T and Manabe T 1982 *IEEE J. Quantum Electron.* **QE-18** 1432–50

Miyashita T and Terunuma Y 1982 *Japan. J. Appl. Phys.* **21** L75–6

Monerie M, Alard F and Maze G 1985 *Electron. Lett.* **21** 1179–81

Moynihan C T, Drexhage M G, Bendow B, Saleh Boulos M, Quinlan K P, Chung K H and Gbogi E 1981 *Mater. Res. Bull.* **16** 25–30

Moynihan C T, Macedo P B, Maklad M S, Mohr R K and Howard R E 1975 *J. Non-Cryst. Solids* **17** 369–85

Nakai T, Mimura Y, Shinbori O and Tokiwa H 1986a *Japan. J. Appl. Phys.* **25** L704–6

Nakai T, Mimura Y, Tokiwa H and Shinbori O 1985a *J. Lightwave Technol.* **LT-3** 565–8

—— 1985b *Electron. Lett.* **21** 625–6

—— 1986b *J. Lightwave Technol.* **LT-4** 87–9

Nassau K 1981 *Bell Syst. Tech. J.* **60** 327–44

Nishii J, Kaite Y and Yamagishi T 1985 *J. Non-Cryst. Solids* **74** 411–5

Ohishi Y, Kanamori T and Mitachi S 1982 *Mater. Res. Bull.* **17** 1563–72

Ohishi Y, Mitachi S, Kanamori T and Manabe T 1983 *Phys. Chem. Glasses* **24** 135–40

Ohishi Y, Mitachi S and Takahashi S 1984a *Mater. Res. Bull.* **19** 673–9

—— 1984b *J. Lightwave Technol.* **LT-2** 593–5

Ohishi Y, Sakaguchi S and Takahashi S 1986 *Electron. Lett.* **22** 1034–5

Ohsawa K and Shibata T 1984 *J. Lightwave Technol.* **LT-2** 602–6

Ohsawa K, Shibata T, Nakamura K and Kimura M 1982 *Tech. Dig. 1st Int. Symp. on Halide and Other Nonoxide Glasses, Cambridge, MA* 23–6

Ohsawa K, Shibata T, Nakamura K and Yoshida S 1981 *Tech. Dig. European Conference on Optical Communication* pp 1.1-1–1.1-4

Olshansky R and Scherer G W 1979 *Proc. 5th ECOC and IOOC, Amsterdam, The Netherlands* pp 12.5.1–12.5.3

Osawa K, Takahashi H, Shibata T and Nakamura K 1980 *Tech. Dig. Annual Meeting of the Institute of Electronics and Communication Engineers of Japan* paper 929 (in Japanese)

Pitt N J, Sapsford G S, Clapp T V, Worthington R and Scott M G 1986 *Proc. Soc. Photo-Optical Instrum.* **618** 124

Poignant H, LeMellot J and Bayon J F 1981 *Electron. Lett.* **17** 295–6

Poignant H, LeMellot J F and Bossis Y 1982 *Tech. Dig. European Conf. on Optical Communication (Cannes)* pp 81–3

Poulain M, Chanthanasinh M and Lucas J 1977 *Mater. Res. Bull.* **12** 151–6

Poulain M, Poulain M, Lucas J and Brun P 1975 *Mater. Res. Bull.* **10** 243–6

Poulain M and Saad M 1984 *J. Lightwave Technol.* **LT–2** 599–602

Robinson M, Pastor R C and Harrington J A 1982 *Tech. Dig. SPIE* (Los Angeles, CA) 320–21

Robinson M, Pastor R C, Turk R R, Devor D P and Braunstein M 1980 *Mater. Res. Bull.* **15** 735–42

Rodney W S, Malitoson J H and King T A 1958 *J. Opt. Soc. Am.* **48** 633–6

Saito M, Masaya T, Sakuragi S and Tanei F 1985 *Appl. Opt.* **24** 2304–8

Saito M and Takizawa M 1986 *J. Appl. Phys.* **59** 1450–2

Sakaguchi S and Takahashi S 1986 *Tech. Dig. 22nd Symp. in Institute of Electrical Communication, Tohoku University, Sendai, Japan* 1–9 (in Japanese)

Sarrach D J and De Neufville J P 1976 *J. Non-Cryst. Solids* **22** 245–67

Savage J A 1965 *Infrared Phys.* **5** 195–204

Savage J A, Webber P J and Pitt A M 1978 *J. Mater. Sci.* **13** 859–64

Schneider H W, Schoberth A, Staudt A and Gerndt C 1986 *Electron. Lett.* **22** 949–50

Shibata S, Kanamori T, Mitachi S and Manabe T 1980a *Mater. Res. Bull.* **15** 129–37

Shibata S, Terunuma Y and Manabe T 1980b *Japan. J. Appl. Phys.* **19** L603–5

Shibata S, Manabe T and Horiguchi M 1981a *Japan. J. Appl. Phys.* **20** L13–6

Shibata S, Terunuma Y and Manabe T 1981b *Mater. Res. Bull.* **16** 703–14

Sugimoto I, Shibuya S, Takahashi H, Kachi S, Kimura M and Yoshida S 1986 *Tech. Dig. 22nd Symp. in Institute of Electrical Communication, Tohoku University, Sendai, Japan* 10–21 (in Japanese)

Takahashi S, Kanamori T, Terunuma Y and Miyashita T 1983 *Tech. Dig. Int. Conf. on Integrated Optics and Optical Fiber Communication (Tokyo)* paper 30A2–4

Takahashi H and Sugimoto I 1984 *J. Lightwave Technol.* **LT-2** 613–5

Takahashi H, Sugimoto I and Sato T 1982 *Electron. Lett.* **18** 398–9

Tokiwa H, Mimura Y, Nakai T and Shinbori O 1985a *Electron. Lett.* **21** 1131–2

Tokiwa H, Mimura Y, Shinbori O and Nakai T 1985b *J. Lightwave Technol.* **LT-3** 569–73

—— 1985c *J. Lightwave Technol.* **LT-3** 574–8

Tran D C 1986 presented at *OFC (Atlanta)*

Tran D C, Burke M J, Sigel G H Jr and Levin K H 1984a *Tech. Dig. Conf. on Optical Fiber Communication (New Orleans, LA)* paper TuG2

Tran D C, Fisher C F and Sigel G H Jr 1982a *Electron. Lett.* **18** 657–8

Tran D C, Ginther R J and Sigel G H Jr 1982b *Mater. Res. Bull.* **17** 1177–84

Tran D C, Ginther R J, Sigel G H Jr and Levin K H 1982c *Tech. Dig. Topical Meeting on Optical Fiber Communication (Phoenix, AZ)* paper TuCC–3

Tran D C, Levin K H, Fisher C F, Burk M J and Sigel G H Jr 1983 *Electron. Lett.* **19** 165–6

Tran D C, Sigel G H Jr and Bendow B 1984b *J. Lightwave Technol.* **LT-2** 566–86

Tregoat D, Liepmann M J, Fonteneau G, Lucas J and Mackenzie J D 1986 *J. Non-Cryst. Solids* **83** 282–96

Van Uitert L G and Wemple S H 1978 *Appl. Phys. Lett.* **33** 57–9

Vasiliev A V, Dianov E M, Plotnichenko V G, Sysoev V K, Bagrov A M, Baikalov P I, Devyatykh G G, Scripachev I V and Churbanov M F 1983 *Electron. Lett.* **19** 589–90

Vechkanov N N, Gur'yanov A A, Devyatykh G G, Dianov E M, Plotnichenko V G, Skripachev I V, Sysoev V K and Vhurbanov M F 1982 *Sov. J. Quantum Electron.* **12** 260–1

Vlasov M A, Devyatykh G G, Dianov E M, Plotnichenko V G, Skripachev I V, Sysoev V K and Churbanov M F 1982 *Sov. J. Quantum Electron.* **12** 932–3

Webber P J 1976 *J. Non-Cryst. Solids* **20** 271–83

Wehr M and Le Sergent C 1986 *Proc. Soc. Photo-Optical Instrum.* **618** 130

Wemple S H 1977 *J. Chem. Phys.* **67** 2151–68

Wood D L, Nassau K and Chadwick D L 1982 *Appl. Opt.* **21** 4276–9
Wood D L and Tauc J 1972 *Phy. Rev.* B **5** 3144–51
Yamane M, Kawazoe H, Inoue S and Maeda K 1985 *Mater. Res. Bull.* **20** 905–11

5 Crystalline Fibers for Infrared Transmission

In this chapter optical fibers made of polycrystals and single crystals are described. TlBr–TlI (KRS-5), silver halides and cesium halides are studied as materials for crystalline fibers which can be used for the transmission of CO_2 and CO laser power.

5.1 Polycrystalline fibers

5.1.1 Introduction

Pinnow *et al* (1978) first prepared TlBr–TlI polycrystalline fibers using an extrusion technique. This was the beginning of research into infrared optical fibers. The main materials studied so far are TlBr–TlI (which is usually called 'KRS-5'), AgCl, and AgCl–AgBr, which have relatively low melting temperatures and high tensile strength.

Polycrystalline fibers show low loss in the wavelength region above $10\,\mu m$. Therefore, the main target for these fibers is use in the power transmission of CO_2 laser light (wavelength $10.6\,\mu m$), which is an important application in the fields of laser surgery and laser welding.

5.1.2 Materials

The best known material used in polycrystalline fiber is TlBr–TlI mixed crystal, which was named KRS-5 during World War II by its German developers. This material has a wide transparent wavelength region of 0.5–$40\,\mu m$, and is almost insoluble in water. Furthermore, it has a moderately high refractive index (2.38). The properties can be adjusted to some extent by varying the proportions of TlBr and TlI. Other properties are listed in table 3.3, together with those of other materials.

Since Pinnow *et al* (1978) announced the fabrication of the TlBr–TlI optical fibers, various research institutes have tried to improve the

146

quality of the TlBr–TlI optical fibers, particularly their light power transmission properties. Sakuragi *et al* (1981) achieved 68 W power transmission of CO_2 laser light. That was the first trial for high power transmission. Then, 130 W transmission was announced by Ikedo *et al* (1986a,b). However, recently a new problem has occurred in transmission loss—loss increase due to the growth of each grain composing the polycrystal. Therefore the suppression of the grain growth must be improved by introducing dopants and by heat treatment.

Other materials for polycrystalline fibers are AgCl, AgBr, and AgCl–AgBr mixed crystal. The properties of these materials are also listed in table 3.3, and are similar to those of TlBr–TlI mixed crystal. The advantage of these materials is that they show a high tensile strength, which is essential for fabricating a fiber by an extrusion technique.

KCl, NaCl, CsI and KBr have also been studied for polycrystalline fibers. These materials show a relatively low transmission loss in their bulk states. The absorption coefficient reaches to as low as 8×10^{-5} cm^{-1} at a 10.6 μm wavelength, which corresponds to 0.035 dB m^{-1}. Moreover, they have a high melting temperature resulting in a stable crystalline state. However, it is difficult to fabricate the polycrystalline fiber by the extrusion technique because of the friction between the alkali halide and the extrusion die (Turk 1982), and so research on the alkali halides has become inactive.

The materials studied so far in various institutes are summarized in table 5.1.

5.1.3 Fabrication methods
5.1.3.1 The extrusion method
Polycrystalline fibers are usually fabricated by an extrusion method.

The fabrication process for TlBr–TlI fibers developed by Sakuragi *et al* (1981) is described here. The starting materials TlBr and TlI are dried in a high vacuum at a temperature around 350 °C and then melted in a nitrogen atmosphere, cast, and sealed into glass ampoules. TlBr and TlI can be also obtained by reacting $TlNO_3$ with HBr and HI. The materials are then crystallized by the vertical Bridgman method. The crystal ingots thus obtained are shaped by mechanical machining to 10 mm in diameter and 30–40 mm in length, ready to be used as fiber preforms. The prepared preforms are finally extruded into fiber form through a diamond-wire die built into an apparatus exerting high pressure on preforms in the container. The apparatus for the extrusion is shown in figure 5.1. The extrusion temperature is varied from 200 to 350 °C, and the extrusion speed is settled at an appropriate rate, typically a few centimeters per minute. The surface of the obtained fiber is smooth enough, except that it carries a few scratches 1–2 μm in width which originate from the microscopic dust trapped between the wire die and the fiber material during extrusion. The grain size of the extruded fiber material is between 10 and 30 μm. The fibers are usually protected by loose-fitting polymer clad, as shown in figure 5.2.

An almost identical fabrication process is used for Ag halides.

Table 5.1 Properties of materials for polycrystalline fibers.

Material	Structure	Fabrication	Property	Reference
TlBr–TlI (KRS-5)	Polymer clad	Extrusion	0.46 dB m^{-1} (10.6 μm)	Pinnow *et al* (1978)
			0.1 dB m^{-1} (10.6 μm)	Harrington (1980)
			0.4 dB m^{-1} (10.6 μm)	Sakuragi *et al* (1981)
	–	Extrusion	0.35 dB m^{-1} (10.6 μm)	Artjushenko *et al* (1984)
	–	Extrusion	0.09 dB m^{-1} (10.6 μm)	Kachi (1984)
	–	Extrusion	0.17 dB m^{-1} (10.6 μm)	Ikedo *et al* (1986a,b)
	Graded index	Diffusion	0.2 dB m^{-1} (10.6 μm)	Sugimoto *et al* (1986)
TlBr	–	–	0.43 dB m^{-1} (10.6 μm)	Harrington (1980)
AgCl	–	Extrusion	6 dB m^{-1} (14 μm)	Chen *et al* (1979)
AgBr	Plastic clad	–	–	Gentile *et al* (1979)
	Plastic clad	–	–	Chen *et al* (1979)
	Teflon clad	Extrusion	0.07 dB m^{-1} (10.6 μm)	Takahashi *et al* (1986)
AgCl–AgBr	–	Extrusion	3.6 dB m^{-1} (10.6 μm)	Garfunkel *et al* (1979)
KCl	Unclad	–	–	Turk (1982)
	–	Extrusion	4.2 dB m^{-1} (10.6 μm)	Harrington (1980)
	–	–	–	Takahashi *et al* (1981)
NaCl	–	–	–	Takahashi *et al* (1981)
KBr	–	–	–	Takahashi *et al* (1981)
CsI	–	–	–	Harrington (1980)

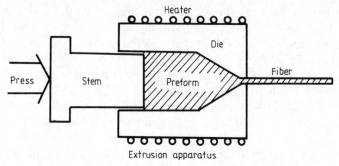

Figure 5.1 The apparatus for the extrusion method.

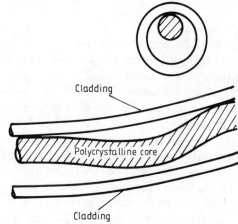

Figure 5.2 Loose polymer cladding for polycrystalline fiber (after Pinnow *et al* 1978).

5.1.3.2 *The rolling method*

When materials such as KCl are extruded through the die, the surface quality of the extruded fibers becomes very poor because of their high solubilities and high melting points. For these reasons an attempt has been made to roll the rod materials repeatedly in order to suppress one-cycle deformation. In this method, the temperature of the roller is maintained at about 300 °C, and the amount of the deformation per cycle is kept at 10%. Turk (1982) obtained a KCl fiber with a diameter of 1 mm.

5.1.3.3 *Fabrication of the clad-type fiber*

In order to stabilize the transmission properties of the polycrystalline fiber, it is necessary to form a cladding surrounding a core. These fibers have been fabricated by Garfunkel *et al* (1979) and Sugimoto *et al* (1986).

TlBr can be used for the cladding material when the core material is TlBr–TlI (KRS-5). In this composition, the difference of the thermal expansion coefficients is very small and the numerical aperture (NA) is more than 0.3. Two fabrication methods have been proposed by Sugimoto *et al* (1986). In the first unclad fibers are extruded by the conventional method, then the extruded fibers are introduced into another die through a guide nipple. Clad-type fibers 1 m long were obtained by this method. However, it is reported that the fabrication of a long fiber is difficult due to the high pressure in the die.

The other method is based on the extrusion of a preform with a core–clad structure. However, the interface between the core and cladding exhibits high irregularity, and so fabrication of a low loss fiber is difficult.

5.1.3.4 Fabrication of the graded-index fiber
Sugimoto *et al* (1986) and Kachi *et al* (1986) have fabricated graded-index polycrystalline fibers. The fabrication process can be divided into two stages. First, a rod composed of TlBr–TlI is inserted into a pipe of TlCl–TlBr. The preform thus fabricated is then extruded into a fiber with a 0.7 mm diameter. The fiber obtained is 3–8 m long. Next, the fiber, which has a step-index distribution, is heated, and thus a gradual compositional variation can be obtained from the diffusion of the anions constituting the fiber. The interface between the core and cladding becomes very smooth after the heat treatment, despite the high irregularity at the interface before the heat treatment. The transmission loss of a fiber thus fabricated was 0.2 dB m^{-1} at a 10.6 μm wavelength. Further loss reduction could be achieved by optimizing the conditions of the heat treatment.

5.1.3.5 Anti-reflection coating at the fiber end face
Anti-reflection coating at the fiber end face is effective in reducing the reflection of the laser light at the fiber input end. The loss due to reflection in the TlBr–TlI fiber reaches as high as 28% because of its high refractive index (2.38). Mouchart (1977) succeeded in reducing the reflection loss to less than 2% by introducing a three-layer anti-reflection coating. Ikedo *et al* (1986a) also used an anti-reflection coating with As_2Se_3 and KCl. Figure 5.3 shows the structure of a coating prepared by evaporation in a vacuum. It consists of three layers, $As_2Se_3/KCl/As_2Se_3$, and is very stable even if the CO_2 laser light with 150 W power is focused on the anti-reflection coating layer.

5.1.4 Properties
5.1.4.1 Transmission loss
The wavelength dependence of transmission loss. Theoretical analysis

shows that the minimum intrinsic loss for crystalline fibers is extremely low. Pinnow *et al* (1978) predicted that halide compounds such as TlBr and TlBr–TlI have a transmission loss of 10^{-2}–10^{-5} dB km^{-1} in the spectral region of more than 4 μm. However, as seen in table 5.1, the minimum loss so far obtained is about 0.1 dB m^{-1}, which is 10^4 times larger than the predicted value. This is mainly due to scattering loss caused by the various scattering centers in the crystals.

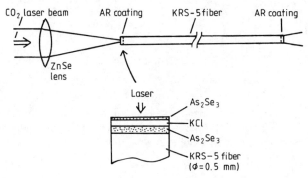

Figure 5.3 The structure of the anti-reflection coating at a fiber end face (after Ikedo *et al* 1986a,b).

The relation between the transmission loss and the wavelength for the TlBr–TlI polycrystalline fibers is shown in figure 5.4 (Sugimoto *et al* 1986). In the figure, the solid line shows the theoretical estimation of the transmission loss, while the dots show the experimental results. As shown, although the estimation shows the minimum loss of less than 10^{-2} dB km^{-1} at a 7 μm wavelength, the minimum loss of the measured fiber is about 100 dB km^{-1}, which takes place in the wavelength region of 20 μm. The large discrepancy between the estimation and the measured value can be explained in terms of the scattering due to both the irregularity at the interface of the core and the crystal defects introduced during the fiber fabrication process (Artjushenko *et al* 1986).

Figure 5.4 also shows that the transmission loss is inversely proportional to the wavelength to the power 2 (Harrington and Sparks 1983). This means that the loss originates from Mie scattering.

Sugimoto *et al* (1986) reported loss reduction by heat treatment of TlBr–TlI fiber. Figure 5.5 shows the transmission loss spectra for various heat-treated TlBr–TlI fibers. The temperature is fixed at 150 °C, and the important parameter is in this case heat-treated time. It can be seen from the figure that the transmission loss is lowered by increasing the heat-treated time. The loss reduction is thought to be due to the reduction in the number of crystal defects (Sugimoto *et al* 1986). It

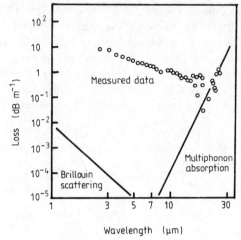

Figure 5.4 The transmission loss spectrum for polycrystalline TlBr–TlI fiber. The solid lines show the theoretical estimate of the transmission loss (after Sugimoto *et al* 1986).

should be noted that the heat treatment is carried out in an inactive gas atmosphere in order to avoid contamination by impurities such as oxygen.

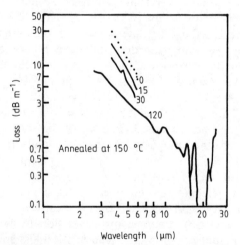

Figure 5.5 Transmission loss spectra for various heat-treated TlBr–TlI fibers. The parameters are heat-treated time at 150 °C (after Sugimoto *et al* 1986).

Harrington and Standlee (1983) discuss the relation between the scattering loss and the residual strain in TlBr–TlI polycrystalline fibers.

In order to clarify the above relation, they measured the total attenuation coefficient due to scattering as a function of stress applied along the fiber axis. They found that the applied stress induces a change in the refractive index, which in turn leads to bulk scattering and excess fiber loss. They also found that the total attenuation coefficient is proportional to the square of the applied stress. From these results, they concluded that the residual stress plays an important role in determining the total attenuation coefficient.

The results so far described in this section are restricted to TlBr–TlI fibers. In silver halide fibers, transmission loss properties are nearly the same as for TlBr–TlI fibers. Taghizadeh *et al* (1984) reported that the background attenuation has a λ^{-2} (λ = wavelength) dependence and that there are strong absorption peaks due to water and silver oxide.

Takahashi *et al* (1986) also analyzed the loss factors in silver halide fibers, and reported that the transmission loss is proportional to $\lambda^{-2.4}$, which is nearly the same as the result reported by Taghizadeh *et al* (1984).

Power transmission. Since Pinnow *et al* (1978) announced a 2 W continuous CO_2 laser beam transmission through a 500 μm diameter fiber, a large number of papers concerning power transmission have been published. The first successful trial of power transmission was done by Sakuragi *et al* (1981). They reported that a TlBr–TlI polycrystalline fiber, 1 mm in diameter and 0.87 m in length, carried laser power of 68 W to its output end and was found to remain free from damage at an incoming laser intensity up to 30 kW cm^{-2}. Recently, Ikedo *et al* (1986a) achieved CO_2 laser power transmission up to 130 W using a TlBr–TlI fiber with 0.1 dB m^{-1} loss. The power density at the input end was as large as 200 kW cm^{-2}. In silver halide fibers, CO_2 laser power of 50 W was transmitted through the fiber with 1 mm diameter (Takahashi *et al* 1986).

Power transmission properties under fiber bending are important in the use of infrared optical fibers. Ikedo *et al* (1986a) studied the change in transmitted power and the aperture of the beam emitted from the output end under fiber bending. Figure 5.6 shows the relation between the transmittance and the bend radius under 90° and 180° bends. The fiber used for the experiment was a TlBr–TlI polycrystalline fiber with 0.5 mm diameter and 1.5 m length. The input power of the CO_2 laser was 20 W. It can be seen from the figure that when the bends are 90° and 180°, the power losses (bending radius: 10 cm) are 4–5% and 8–10% respectively. Furthermore, although the aperture of the emitted beam is 10–13° for the straight fiber, it increases by 3–4° and 8–10° for 90° and 180° bend, respectively. Therefore, these changes in properties on bending must be considered when the optics for practical systems are designed.

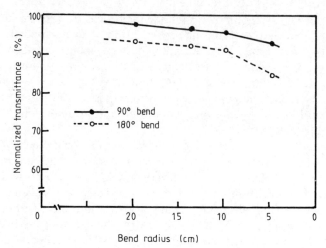

Figure 5.6 The relation between transmittance and bend radius under 90° and 180° bend in TlBr–TlI polycrystalline fibers (0.5∅ × 1500 (mm), CO_2 laser output power = 20 W) (after Ikedo *et al* 1986a,b).

Anti-reflection coating at the fiber end faces is also particularly important for power transmission. Mouchart (1977) and Ikedo *et al* (1986a) reported anti-reflection coating techniques. A coating of $As_2Se_3/KCl/As_2Se_3$ is effective for CO_2 laser light transmission (Ikedo *et al* 1986a). The lifetimes were 1000 h and 10 h for 40 W and 100 W input power, respectively. Lifetime here is defined as the time at which the transmittance decreases to 90% of the initial value. No noticeable degradation could be observed at the end face of the fiber, except that the third coating layer showed slight degradation.

Degradation of transmission properties. The most important problem to be solved in the polycrystalline fibers may be loss increase incurred gradually over a long period of time.

Sugimoto *et al* (1986) studied the relation between the loss increase of a TlBr–TlI fiber and the amount of water in the atmosphere. A noticeable loss increase was observed for a high content of water vapor, while there was almost no increase in loss under dry air. The fiber studied showed no new absorption peaks and no changes in the wavelength dependence of the loss. This means that the loss increase is governed by scattering and that the absorption due to various impurities does not contribute to the loss increase. Sugimoto *et al* (1986) suggested that the loss increase is due to the corrosion caused by water which occurs at the surface of the core. Furthermore, they suggested that the loss increase can be suppressed by heat treatment. The heat-treated fiber showed almost no increase in transmission loss even under water vapor. This is due to the improvement of the crystallinity caused by the heat treatment.

Another important problem in the polycrystalline fiber is the loss increase due to plastic deformation caused by repeated bending. Takahashi *et al* (1986) reported that the transmittance of silver halide polycrystalline fiber decreases to 80% of the initial value after bending 60 000 times (the fiber diameter is 0.7 mm and the bending radius is 25 cm).

5.1.4.2 Refractive index and dispersion
The refractive indices of typical polycrystalline materials used for infrared optical fibers are listed in table 3.3. As shown, the refractive index of TlBr–TlI is relatively large while that of KCl is almost the same as that of silica glass, which is used widely for conventional optical fibers.

The zero material dispersion wavelength was identified for a wide range of oxides, fluorides, chlorides, and bromides by Nassau (1981), including halide crystals for infrared optical fibers. The zero material dispersion wavelengths calculated theoretically for typical halide fiber materials are located at a considerably longer wavelength than those for oxide, fluoride, and chalcogenide glasses (Nassau 1980). The material dispersion slope at zero dispersion wavelength can also be calculated, and results indicates that halide crystals such as TlBr–TlI and CsI show much smaller slopes than oxide, chalcogenide and fluoride glasses.

5.1.4.3 Mechanical and thermal properties
The mechanical and thermal properties of AgBr, TlBr–TlI and CsBr are listed in table 3.3 (Takahashi *et al* 1986).

It should be noted that grain size plays an important role in improving the fiber strength in polycrystalline fibers. Garfunkel *et al* (1979) discussed the strength of silver halide material. They showed that although the grains in pure materials usually grow rather rapidly, alloying of halide materials can reduce grain growth and increase yield strength without increasing optical absorption. Alloying of KCl with RbCl and of AgCl with AgBr is effective in increasing the yield strength.

5.1.5 Summary
The materials used for polycrystalline fibers are thallium halides and silver halides, which are characterized by their ductile properties. These fibres are fabricated by an extrusion method.

The main transmission loss of the fiber is Mie scattering loss, which has a wavelength dependence of roughly λ^{-2}. Therefore the transmission loss is low in the wavelength region of more than 10 μm, which is useful for transmitting CO_2 laser power (10.6 μm wavelength). A loss of as low as 0.1 dB m^{-1} at a 10.6 μm wavelength has been achieved in various research institutes. Furthermore, the CO_2 laser power exceeding 130 W has been successfully transmitted through polycrystalline fiber.

However, it should be noted that transmission loss is increased by plastic deformation and grain growth. In particular, the water in the atmosphere is found to increase the transmission loss because of the corrosion caused by the water. Heat treatment and doping techniques may be effective in suppressing the loss increase.

5.2 Single-crystalline fibers

5.2.1 Introduction
The fabrication of crystalline fibers is of interest since crystals such as alkali halides and thallium halides have wide transparent windows as well as low intrinsic transmission losses. The first fibers to be fabricated were polycrystalline fibers, which were extruded by dies. However, polycrystalline fibers show a scattering loss due to the existence of grain. Since this scattering loss increases towards shorter wavelengths, these fibers can be used only in the longer wavelength region, such as for the CO_2 laser wavelength. To avoid scattering loss, single-crystalline fibers have been fabricated.

Single-crystalline fibers made of halide compounds were first fabricated by Bridges *et al* (1980). They fabricated an AgBr fiber with a loss of 8 dB m^{-1} (10.6 μm wavelength). Following this fabrication, a variety of halide compounds such as TlBr–TlI, KCl, CsBr and CsI were studied for the fabrication of the single-crystalline fibers.

The advantage of these single-crystalline fibers is that they possess a wide transparent wavelength region from visible to far-infrared. This is very convenient for power transmission because two kinds of light can be used; one, the infrared laser light, is for power transmission, and the other, visible light, is for the alignment of the laser beam.

5.2.2 Materials
The materials used for single-crystalline fibers are basically the same as those used for the polycrystalline fibers mentioned in the previous section. The materials so far studied are listed in table 5.2. Halide compounds such as KCl, CsBr, CsI, AgBr and TlBr–TlI have been studied for the fiber fabrication. Among them, potassium halides such as KCl show poor transmission characteristics because of their NaCl-type crystal structures. However, the other materials have been used successfully for infrared optical fibers. In particular, CsBr and CsI are

advantageous because they are non-toxic and show none of the photo-sensitivity shown by the silver halides (Mimura *et al* 1981). The properties of these materials are shown in table 3.3.

Table 5.2 Properties of single crystalline fibers.

Material	Structure	Fabrication	Property	Reference
TlBr–TlI (KRS-5)	Unclad	Modified pulling down (MPD)	–	Mimura *et al* (1980)
AgBr	Unclad	Pressure applying	2×10^{-2} cm^{-1} (10.6 μm)	Bridges *et al* (1980)
KCl	Unclad	Growth in vapor space over solution	0.5 cm^{-1} (10.6 μm)	Tangonan *et al* (1973)
CsBr	Unclad	MPD	5 dB m^{-1} (10.6 μm)	Mimura *et al* (1981)
			0.3–0.4 dB m^{-1} (10.6 μm)	Mimura and Ota (1982)
CsI	Plastic loose clad	MPD	13 dB m^{-1} (10.6 μm)	Okamura *et al* (1980)
Al$_2$O$_3$(Cr)	Sapphire clad	Laser melting	–	Burrus and Coldren (1977)
	Unclad	Laser melting	7.4 dB m^{-1} (1.06 μm)	Fejer *et al* (1984)

The most important feature of the single-crystalline fiber is the loss increase due to plastic deformation caused by repeated bending. Mimura and Ota (1982) and Mimura *et al* (1982) reported that fiber losses increased after two hundred bending repetitions when the radius of the bending curvature was as small as 5 cm. This plastic deformation is attributed to the slip traces which appeared on the fiber surface. To avoid loss increase the fiber should be handled carefully, for example by restricting the bending curvature.

The weak absorption tail discussed in the following section is also important in determining the transmission loss of single-crystalline fibers. A weak absorption tail is thought to be intrinsic to the alkali halide crystals, and so a method of lowering the weak absorption tail is required for ultra-low loss. However, there would be no problem if only short haul transmission were considered.

Materials other than halide compounds have also been studied for fabricating single-crystalline fibers. Al_2O_3, Nd:YAG and $LiNbO_3$ have been studied mainly for use in fiber-type laser rods (Burrus and Coldren 1977, Fejer *et al* 1984). These fibers can also be used for infrared light transmission. Al_2O_3 fiber was reported to have a relatively low loss of 7.4 dB m^{-1} at 1.06 μm wavelength (Fejer *et al* 1984). Single-crystalline fibers composed of Ge have also been fabricated (Nakagawa 1964). Although this Ge fiber was fabricated for electrical transmission, its use as an infrared optical fiber is interesting because of its intrinsic high transparency in the infrared region.

5.2.3 Fabrication methods
5.2.3.1 The edge-defined film-fed growth method
The edge-defined film-fed growth method (EFG) has been developed mainly for the fabrication of filamentary sapphire (LaBelle and Mlavsky 1971, Pollock 1972). This method is also attractive for the fabrication of infrared optical fibers.

The EFG method uses a capillary phenomenon. The filamentary crystal can be grown from the die. Figure 5.7 shows a schematic diagram of the top part of the crucible and the Mo growth orifice (die) showing a sapphire filament being withdrawn from a thin liquid meniscus pool at the top of the growth orifice (Pollock 1972). The pool is continuously fed by capillary action from the melt held in the Mo crucible. The filament thus obtained had a nominal 250 μm diameter and was grown at a rate in the range 2.5 to 7.5 cm min^{-1}.

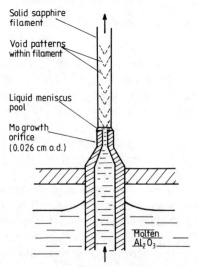

Figure 5.7 A schematic diagram showing the Mo growth orifice and the sapphire filament being withdrawn from the molten meniscus pool (after Pollock 1972).

The most important factors in fabricating a fiber by the EFG method are the following. Firstly, the choice of die material should be appropriate to the fiber material. In order to obtain a smooth fiber surface, the die must be dampened by the melt. Secondly, precise control of temperature is needed to maintain a constant fiber diameter. However, it is, in general, very difficult to satisfy these conditions in fabricating the infrared fiber. Therefore, only a few trials have been reported on fiber fabrication by the EFG method.

5.2.3.2 The pressure-applying growth method

This method has been developed by Bridges *et al* (1980) for the fabrication of AgBr single-crystalline fibers. In this method, the melt is supplied from the die by gas pressure instead of by capillary action. Figure 5.8 shows a schematic of the apparatus. Fiber growth occurs close to the exit of a nozzle that determines the size and cross section of the crystal. The nozzle terminates one arm of a fused-quartz U tube containing the liquid charge, which is kept molten by a surrounding oven. The rate of liquid feed to the nozzle can be sensitively controlled by N_2 gas pressure applied to the other arm of the U tube. A small oven around the nozzle tip with an independently adjustable temperature and a water-cooled element which can be placed where required enable the interface to be accurately positioned. It is possible to grow fibers of a size determined by both the outside and inside diameters of the nozzle.

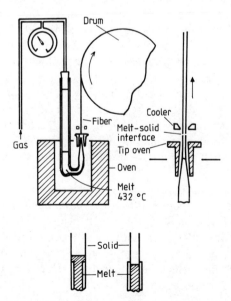

Figure 5.8 A schematic of the pressure-applying growth method (after Bridges *et al* 1980).

Smooth, clear fibers with diameters between 0.35 and 0.75 mm have been grown at rates up to 2 cm min^{-1}. Bridges *et al* (1980) insisted that a much higher growth rate is possible and that the allowable growth rate is inversely proportional to the cross-sectional area of the crystal. Stable crystal growth is indicated with the [100] direction along the fiber axis. This direction appears to be the most favorable and is not affected by the type or orientation of the original seed. The transmission loss of the fiber thus obtained is 2×10^{-2} cm^{-1} at 10.6 μm wavelength.

5.2.3.3 The modified pulling-down method

This method is essentially based on a floating-zone technique rather than on an EFG process. Mimura *et al* (1980) developed the method and obtained fibers of various materials, including TlBr–TlI and CsBr. A schematic diagram of the procedure is shown in figure 5.9. In this system the crucible is constructed in three parts: a melt container in which the raw material rod is melted by an upper side heater; a capillary, the diameter of which is chosen to give the required flow rate, through which the melt flows into a shaper; and the shaper which controls the cross-sectional profile of the product. One of the merits of this crucible design is that one can select either wetted or unwetted materials for the shaper. Separation between the shaper and the container (the length of the capillary) is determined so that heat flux from the upper side heater is low enough at the bottom of the shaper. The diameter of the lower side heater is designed to be small, corresponding to the diameter of the shaper. These designs make it possible to provide a steep thermal gradient at the bottom edge of the shaper so that a high growth rate can be achieved. Incidentally, in the EFG process it is difficult to achieve such a steep thermal gradient at the top of the die, because the size of the heater is restricted by the diameter of the crucible.

Using this technique, TlBr–TlI fibers with a diameter of 0.6–1 mm and a length of 1–2 m have been grown at the high growth rate of 0.5–3 cm min^{-1}. Mimura *et al* (1980) have suggested that this technique can be used for any material, including NaCl, KCl, KBr, LiF, CaF$_2$, BaF$_2$, etc., for which nonreactive crucible materials can be easily found.

5.2.3.4 The laser melting method

Single-crystalline Cr-doped Al$_2$O$_3$ fibers have been grown by a floating-zone technique from small source rods locally melted by a CO$_2$ laser (Burrus and Coldren 1977). In addition, by using the same laser to melt only the surface of the fiber and then allowing it to regrow, the Cr concentration in the surface layer can be reduced by a factor of 100 or more by selective evaporation.

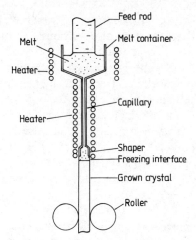

Figure 5.9 A schematic of the modified pulling-down method (after Mimura *et al* 1980).

Figure 5.10 is a schematic of the growth apparatus. Illustrated is the primary growth step in which a uniform *c*-axis ruby fiber roughly 250 μm in diameter is grown from a source rod roughly 1 mm in diameter at rates which vary from 0.5 to 2 cm min^{-1}. All operations are in air. The source rods consist either of ceramics formed from mixed sintered Al_2O_3 and Cr_2O_3 powders, or of lengthwise-slotted single-crystal bars of Al_2O_3 with Cr_2O_3 loaded into the slot.

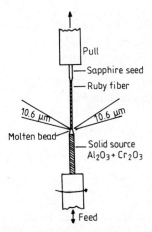

Figure 5.10 A schematic of the laser melting method (after Burrus and Coldren 1977).

The second step of the procedure, in which a layer on the fiber surface is melted and regrown, is also accomplished using the apparatus depicted in figure 5.10, except that the upper pulling head is removed and the output of the CO_2 laser is reduced in power and aimed, slightly defocused, at the fiber surface. The fiber then is rotated and lowered slowly so that a molten spot at each laser focal point is moved along the length of the fiber in an overlapping spiral path. The extent of the Cr evaporation is determined by the time the surface is molten, i.e., by the speed of rotation and feed, and/or the number of passes through the beam. Green Cr_2O_3 powder is deposited on nearby cool surfaces during this operation. Using this technique, single-crystalline sapphire-clad ruby-core fibers have been obtained.

Recently, Fejer *et al* (1984) developed the laser melting method by introducing optical, mechanical and electronic control systems. Single-crystalline fibers of four refractory oxide materials, Al_2O_3, $Cr:Al_2O_3$, Nd:YAG, and $LiNbO_3$ were grown. The diameter of the fiber can be controlled to be as small as 20 μm. Measured optical loss at 1.06 μm for a 10 cm long, 170 μm diameter $Cr:Al_2O_3$ fiber was 7.4 dB m^{-1}.

5.2.3.5 The pulling method

Nakagawa (1964) proposed a method which enables us to fabricate single-crystalline Ge fibers. The fabrication procedure is described in figure 5.11. In this method, Ge ingots are inserted into glass tubes and then are heated by a burner ((a) and (b)). Then the melted Ge, together with the surrounding glass, is pulled down from a nozzle in the crucible ((c) and (d)). The Ge fiber is obtained by removing the surrounding glass by etching ((e) and (f)). It should be noted that if the melting temperature of the glass is selected to be below that of the Ge, the fracture of the surrounding glass due to the mismatch of the thermal expansion coefficients can be prevented.

The factors determining the diameter of the Ge fiber are:

(i) the ratio of the quantity of the Ge ingot to that of the glass tube,
(ii) the viscosity of the Ge and glass,
(iii) the pulling speed,
(iv) the inner diameter of the nozzle.

The axial direction of the fiber is inclined at 5–10° to the [100] direction of the crystal.

Nakagawa (1964) has tried to construct a thermistor, strain gauge, and microelectric circuits using this Ge fiber. The fiber can, however, also be used for infrared light transmission, since Ge has a wide transparent wavelength region up to more than 10 μm (Matsumura and Katsuyama 1980).

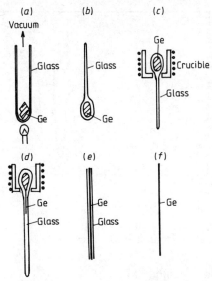

Figure 5.11 A schematic of the procedure for the pulling method (after Nakagawa 1964).

5.2.4 *Properties*
5.2.4.1 *Transmission loss of halide fiber*
The wavelength dependence of transmission loss. The most important advantage of single-crystalline fibers is that they show no scattering loss due to the existence of grain, which is the main transmission loss in polycrystalline fibers. However, in single-crystalline fibers it is difficult to completely eliminate the outer radius variation, which also causes the scattering loss. In particular, alkali halide fibers such as KCl and KBr fibers tend to show large surface scattering.

The wavelength dependence of the single-crystalline fiber has been studied in detail by Mimura (Mimura *et al* 1982, Mimura 1983). Figure 5.12 shows the transmission loss spectrum for a CsBr single-crystalline fiber. The fiber is fabricated by a modified pulling-down method and the fiber length is 1 m. The minimum loss of this fiber is below $0.1 \, \mathrm{dB \, m^{-1}}$ at a $5.7 \, \mu\mathrm{m}$ wavelength. The strong absorption bands are due to impurity ions of SO_4^{2-}, CO_3^{2-} and NO_2^{-}. It can be seen that in the infrared wavelength range the absorptions due to these ions are much stronger than those of transition-metal impurities. The tail of the transmission loss appearing in the wavelength region between 0.5 and $5.7 \, \mu\mathrm{m}$ is due to scattering. The transmission loss at $10.6 \, \mu\mathrm{m}$ is $0.4 \, \mathrm{dB \, m^{-1}}$, which is thought to be mainly caused by impurity absorption. The value of $0.4 \, \mathrm{dB \, m^{-1}}$ is the lowest one attained at a $10.6 \, \mu\mathrm{m}$

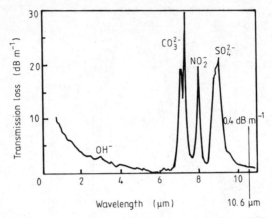

Figure 5.12 The transmission loss spectrum for CsBr single-crystalline fiber (after Mimura 1983).

wavelength by single-crystalline fibers.

Mori and Izawa (1980) reported an interesting result on the transmission loss of infrared material. They measured the absorption coefficients of LiF, BaF_2, CsI, KCl, TlBr–TlI and SiO_2 glass by laser calorimetry. The materials are all crystalline forms, except SiO_2. Their results are shown in figure 5.13, where the solid lines represent the data reported before their measurement. It can be seen that the TlCl–TlBr absorption coefficients have a kink and some absorption tail is observed below $1.5 \times 10^{-1} \ cm^{-1}$ in absorption coefficient. Other materials have similar characteristics in the visible wavelength region. These measured absorption tails can be expressed by

$$\alpha = \alpha_0 \exp(\hbar\omega/E), \qquad (5.1)$$

where α is the absorption coefficient, $\hbar\omega$ is the photon energy, and α_0 and E are constants. Slope factors E for the absorption tails have almost the same value in the range 0.3–0.5 for both crystalline and amorphous materials, except for BaF_2.

These absorption tails are called 'weak absorption tails' (Wood and Tauc 1972), and are attributed to absorption by defects, disorders, and impurities. It should be noted however that the weak absorption tail can be found even in high purity materials, such as a zone-refined NaCl crystal and silica glass for optical fiber, and is therefore thought to result from intrinsic defects in thermal equilibrium.

It can also be seen from figure 5.13 that the total loss in the wavelength region 0.4–2 μm depends mainly on the weak absorption tail, because this slope is flatter than that of multiphonon absorption. Therefore, Mori and Izawa (1980) insisted that the weak absorption tail plays an important role in determining the transparency limitation for low loss materials, particularly materials for infrared optical fibers.

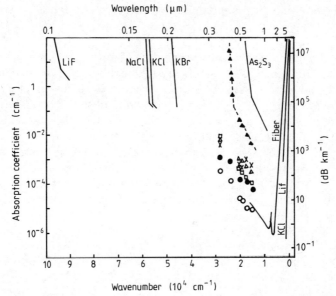

Figure 5.13 Measured absorption coefficients versus wavenumber for LiF, KCl, CsI, BaF$_2$, KRS-6 (TlCl–TlBr) and SiO2 (▲, KRS-6; ×, BaF$_2$; △, KCl; □, CsI; ◆, LiF; ○, SiO$_2$ (Suprasil-W1), (after Mori and Izawa 1980).

Power transmission. Mimura and Ota (1982) reported CO$_2$ laser light transmission through a CsBr single-crystalline fiber. Four samples of fibers were prepared for optical measurements. Both ends of the fibers were polished by abrasive papers, and then inserted into Teflon tubes for loose cladding. Experiments of high power transmission were performed using a CO$_2$ laser with the maximum output power of 55 W. A ZnSe lens was used in focusing the laser beam. In this experiment, both ends of the Teflon tubes were jacketed by thin aluminum sheets to protect them from thermal damage.

Three fibers with losses of 0.3–0.4 dB m^{-1} could transmit a continuous laser beam above 40 W for 30 min without any degradation. On the other hand, in the case of the fiber with the loss of 1.2 dB m^{-1}, the output end was molten after 2.2 min transmission. The maximum output power was 47 W in the 1 mm diameter fiber, which corresponds to the power density of 6 kW cm^{-2}.

Note that the connecting losses of these fibers are much smaller than those of polycrystalline TlBr–TlI fibers. The estimated loss by reflection is about 11% for CsBr fibers but about 27% for the TlBr–TlI fibers. As a result of the relatively low connecting losses, CsBr fibers can favorably transmit more than 70% laser power per meter without surface coating.

In addition to the above results, it has also been reported that the bending losses are barely observable so long as the radius of curvature is above 5 cm, although some bending losses occur at a radius of curvature of 1 cm.

Degradation of transmission properties. Figure 5.14 shows the loss increase caused in a CsBr single-crystalline fiber by repeated bending (Mimura and Ota 1982). As shown, when the radius of curvature is 14 cm, degradation of the fiber loss is not observed even after one thousand bending cycles. However, the fiber loss increases after two hundred bending cycles when the radius of curvature is 5 cm. This increase of the fiber loss due to bending is caused by the increase in scattering loss resulting from slip traces which appear on the fiber surface. Care must therefore be taken when handling the fiber in order to prevent the loss increase by repeated bending.

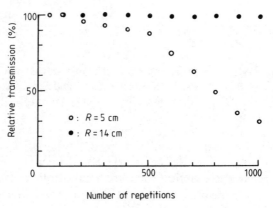

Figure 5.14 The degradation of 1.6 mm diameter fibers by repeated bending for two different bending radii R (after Mimura *et al* 1982).

5.2.4.2 Transmission loss of other fibers
Only a few papers describing the transmission losses of single-crystalline fibers other than halide fibers have been published to date. Fejer *et al* (1984) reported the transmission loss of a Cr-doped Al_2O_3 single-crystalline fiber fabricated by a laser melting method. A 10 cm long, 170 μm diameter fiber showed 72% transmission of an incident 1.06 μm Nd:YAG laser beam. Taking into account the Fresnel reflection losses from the fiber end faces, the transmission loss can be calculated to be 1.7% cm^{-1} or, equivalently, 0.074 dB cm^{-1}.

5.2.4.3 Refractive index and dispersion
The refractive indices of TlBr–TlI, AgBr, KCl, CsBr and CsI materials

studied for single-crystalline fibers are listed in table 3.3. In general, the refractive indices of the alkali halide fibers such as CsBr and KCl fibers are smaller than that of TlBr–TlI fiber. The connecting loss of alkali halide fiber is therefore much smaller than that of TlBr–TlI fiber because of its small Fresnel reflection.

The refractive index properties of the other single-crystalline fibers such as Al_2O_3, $LiNbO_3$ and Ge fibers are also listed in table 3.3.

5.2.4.4 Mechanical and thermal properties

These properties have already been discussed in the section on polycrystalline fibers.

The thermal and mechanical properties of Al_2O_3 and Ge are listed in table 3.3, together with the values for refractive indices.

5.2.5 Summary

Materials for single-crystalline fibers are almost the same as those for polycrystalline fibers. TlBr–TlI, AgBr, KCl, CsBr, and CaI have mainly been studied, and various fabrication methods, such as edge-defined film-fed growth, pressure-applying growth, and the modified pulling-down method have been proposed. Transmission loss of $0.3 \, \mathrm{dB \, m^{-1}}$ was obtained at a $10.6 \, \mu m$ wavelength by using a CsBr fiber (Mimura and Ota 1982). A 47 W continuous laser light was transmitted through a 1 mm diameter fiber.

The advantage of the single-crystalline fibers is that they possess a wide transparency region from visible to far-infrared. This makes it possible to transmit a visible guiding light as well as high-power infrared light. On the other hand, the disadvantage is that there is a significant loss increase due to the plastic deformation caused by repeated bending. Therefore these fibers must be handled carefully, for example, by introducing a method to prevent a high bending curvature.

Single-crystalline fibers made of Al_2O_3 and Ge are also important in transmitting infrared light. These fibers can also be used for a variety of applications as in micro-laser rods and thin semiconductor elements.

References

Artjushenko V G, Butvina L N, Vojtsekhovsky V V, Dianov E M and Kolesnikov J G 1986 *J. Lightwave Technol.* **LT-4** 461–5
Artjushenko V G, Butvina L N, Vojtsehouskii V V, Dianov E M and Prokhorov A M 1984 *Electron. Lett.* **20** 93–4
Bridges T J, Hasiak J S and Strand A R 1980 *Opt. Lett.* **5** 85–6
Burrus C A and Coldren L A 1977 *Appl. Phys. Lett.* **31** 383–4
Chen D, Skogman R, Bernal E G and Butter C 1979 *Fiber Optics* ed B Bendow

and S S Mitra (New York: Plenum) pp 119–22

Fejer M M, Nightingale J L, Magel G A and Byer R L 1984 *Rev. Sci. Instrum.* **55** 1791–6

Garfunkel J H, Skogman R A and Walterson R A 1979 *IEEE/OSA Conf. on Laser Eng. and Appl.* paper 8.1

Gentile A L, Braustein M, Pinnow D A, Harrington J A, Hobrock L M, Mayer J, Paster R C and Turk R R 1979 *Fiber Optics* ed B Bendow and S S Mitra (New York: Plenum) pp 105–18

Harrington J A 1980 *Proc. Soc. Photo-Optical Instrum.* **227** 133–7

Harrington J A and Sparks M 1983 *Opt. Lett.* **8** 223–5

Harrington J A and Standlee A G 1983 *Appl. Opt.* **22** 3073–8

Ikedo M, Watari M, Tateishi F, Fukui T and Ishiwatari H 1986a *Tech. Dig. 22nd Symposium in Institute of Electrical Communication, Tohoku University, Sendai, Japan* 22–9 (in Japanese)

Ikedo M, Watari M, Tateishi F and Ishiwatari H 1986b *J. Appl. Phys.* **60** 3035–9

Kachi S 1984 *Tech. Digest 45th Fall Meeting of Japanese Society of Applied Physics* p 50 (in Japanese)

Kachi S, Kimura M and Shiroyama K 1986 *Electron. Lett.* **22** 230–1

LaBelle H E Jr and Mlavsky A I 1971 *Mater. Res. Bull.* **6** 571–80

Matsumura H and Katsuyama T 1980 *Japanese patent Tokkaisho* 57–136606

Mimura Y 1983 *Oyobutsuri (Applied Physics)* **52** 311–12 (in Japanese)

Mimura Y, Okamura Y, Komazawa Y and Ota C 1980 *Japan. J. Appl. Phys.* **19** L269–72

—— 1981 *Japan. J. Appl. Phys.* **20** L17–18

Mimura Y, Okamura Y and Ota C 1982 *J. Appl. Phys.* **53** 5491–7

Mimura Y and Ota C 1982 *Appl. Phys. Lett.* **40** 773–5

Mori H and Izawa T 1980 *J. Appl. Phys.* **51** 2270–1

Mouchart J 1977 *Appl. Opt.* **16** 2722–8

Nakagawa T 1964 *J. Institute of Electrical Engineering of Japan* **84** 1614–18 (in Japanese)

Nassau K 1980 *Electron. Lett.* **16** 924–5

—— 1981 *Bell Syst. Tech. J.* **60** 327–37

Okamura Y, Mimura Y, Komazawa Y and Ota C 1980 *Japan. J. Appl. Phys.* **19** L649–51

Pinnow D A, Gentile A L, Standlee A G, Timper A J and Hobrock L M 1978 *Appl. Phys. Lett.* **33** 28–9

Pollock J T A 1972 *J. Mater. Sci.* **7** 631–48

Sakuragi S, Saito M, Kubo Y, Imagawa K, Kotani H, Morikawa T and Shimada J 1981 *Opt. Lett.* **6** 629–31

Sugimoto I, Shibuya S, Takahashi H, Kachi S, Kimura M and Yoshida S 1986 *Tech. Dig. 22nd Symp. in Institute of Electrical Communication, Tohoku University, Sendai, Japan* 10–21 (in Japanese)

Taghizadeh M R, Barton J S, Melling P J, Middleton R J and Smith S D 1984 *Opt. Commun.* **51** 171–4

Takahashi K, Murakami K and Yokota M 1981 *Tech. Dig. Spring Meeting of Japan. Soc. Appl. Phys.* p 225 (in Japanese)

Takahashi K, Yoshida N and Yamauchi K 1986 *Tech. Dig. 22nd Symp. in Institute of Electrical Communication, Tohoku University, Sendai, Japan* 30–5 (in Japanese)

Tangonan G, Paster A C and Paster R C 1973 *Appl. Opt.* **12** 1110–11

Turk R R 1982 *Advances in IR Fibers, Tech. Dig. Soc. Photo-Optical Instrum., Los Angeles, CA* paper 320–17

Wood D L and Tauc J 1972 *Phys. Rev.* B **5** 3144–50

6   Hollow Waveguides for Infrared Transmission

This chapter describes hollow waveguides, i.e., metallic hollow waveguides and dielectric hollow waveguides composed mainly of glass materials. Some new concepts are required for understanding the light guiding characteristics of these waveguides because they differ slightly from those of the optical fibers described in the previous chapter. These waveguides have been studied mainly for laser power transmission, particularly CO_2 laser transmission at 10.6 μm wavelength.

6.1 Metallic hollow waveguides

6.1.1 Introduction
Metallic hollow waveguides have been studied mainly for infrared light power transmission, particularly CO_2 laser power transmission at 10.6 μm wavelength. They were first proposed by Nishihara et al (1974). Garmire et al (1976b) then studied extensively the transmission properties of the waveguides, which were based on the hollow waveguides proposed by Marcatili and Schmeltzer (1964). The waveguides studied were parallel-plate metallic waveguides fabricated by two metallic plates. Recently, Kubo and Hashishin (1986) succeeded in constructing a machine for laser surgery which uses a parallel-plate metallic waveguide.

Circular metallic hollow waveguides have also been studied. In particular, Miyagi et al (1983a) developed dielectric-coated metallic hollow waveguides whose transmission properties can be improved by introducing dielectric coatings on the inner surfaces.

Metallic hollow waveguides can, in general, transmit higher laser power than optical fibers whose cores are made of various dielectric materials. However, they are not as flexible as conventional optical fibers, and so their application is rather restricted. Further improve-

170

ments in the mechanical properties to give some flexibility are still needed for laser power transmission.

6.1.2 Parallel-plate waveguides
6.1.2.1 Basic structure
The basic structure of parallel-plate metallic waveguides is shown in figure 6.1. Two strips of aluminum sheet are cut to form the top and bottom walls, and these are separated by brass shimstocks whose edges form the side walls. Furthermore, as shown in the figure, these waveguides have a function allowing the bends and twists needed in practical use.

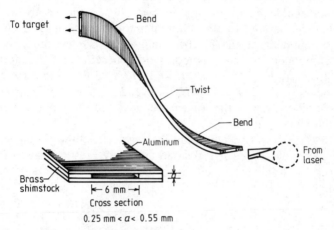

Figure 6.1 The basic structure of a parallel-plate metallic waveguide (after Garmire *et al* 1980, © 1980 IEEE).

The light propagation characteristics of parallel-plate metallic waveguides have been studied extensively by Garmire *et al* (1980). In this section, the formulation of propagation characteristics discussed by Garmire *et al* (1980) is shown. In principle, parallel-plate waveguides are equivalent to planar waveguides when the width of the hollow part is much greater than the height. However, they act as rectangular waveguides when the sidewalls are brought close enough together to interact with the incident laser beam.

The loss of a waveguide can be expressed by using the power loss coefficient

$$\alpha = \alpha^{\text{TE}} + \alpha^{\text{TM}} + \alpha_{\text{R}} + \alpha_{\text{T}} \tag{6.1}$$

where α^{TE} is the loss coefficient of the transverse electric mode and α^{TM} is that of the transverse magnetic mode, α_{R} is the additional loss caused by the bend, and α_T is the twist loss. The power loss coefficient α (which is sometimes simply called the 'loss coefficient') is defined by equation (2.18). Therefore, the power loss coefficient is equivalent to the absorption coefficient. Expressions for all the terms in equation (6.1) can be derived theoretically except for α_R, the additional loss due to the bend. However, the total loss in a bent guide, α_B, rather than the loss added by a bend, α_R, can be calculated. Each term is described in the following.

Straight waveguide losses. When the sidewalls are spaced about 1 cm apart, the laser beam does not diffract appreciably at the sidewalls. This arrangement is thus equivalent to TE mode propagation in a planar (slab) waveguide having aluminum walls when the polarization of the input light is parallel to the aluminum walls. The TE mode has only a y-axis component of the electric field while the TM mode has only a y-axis component of the magnetic field. Therefore the TE mode corresponds to $E_x = E_z = 0$, $H_y = 0$, and $H_z \neq 0$, while the TM mode has components of $H_x = H_z = 0$, $E_y = 0$, and $E_z \neq 0$. The z-axis direction here is the direction of the waveguide axis, and the x-axis is perpendicular to the aluminum walls.

The loss analysis of each mode in a perfect hollow planar waveguide, whose walls have specular reflection and obey Fresnel's laws of reflectivity, gives the following results for power loss coefficients of TE and TM mode light:

$$\alpha_m^{TE} = \frac{m^2\lambda^2}{a^3} Re[(\tilde{n}^2 - 1)^{-1/2}] \simeq \frac{m^2\lambda^2}{a^3} Re(1/\tilde{n}) \qquad (6.2a)$$

$$\alpha_m^{TM} = \frac{m^2\lambda^2}{a^3} Re[\tilde{n}^2(\tilde{n}^2 - 1)^{-1/2}] \simeq \frac{m^2\lambda^2}{a^3} Re(\tilde{n}), \qquad (6.2b)$$

where a is the height of the waveguide, λ is the wavelength, and $\tilde{n}$ is the complex refractive index of the walls. m is the mode number. α_m^{TE} and α_m^{TM} are the power loss coefficients of mth TE modes and mth TM modes, respectively. These equations can also be derived very simply using ray optics and the single incident reflection properties of the waveguide walls.

It will be useful to express the waveguide loss in terms other than of the complex refractive index. Equation (6.2) can be expressed in terms of the slope (S^{TE} and S^{TM} of the single incidence reflectivity (or absorption) near grazing incidence:

$$S^{TE} = \frac{\partial R^{TE}}{\partial \theta}\bigg|_{\theta=0} = 4Re(1/\tilde{n}) \qquad (6.3a)$$

$$S^{TM} = \frac{\partial R^{TM}}{\partial \theta}\bigg|_{\theta=0} = 4Re(\tilde{n}), \qquad (6.3b)$$

where R^{TE} and R^{TM} represent the reflectivities for TE and TM modes, respectively, and θ is the reflection angle. This means that the straight planar guide loss can be expressed as follows:

$$\alpha_m^{TE} = \frac{m^2\lambda^2}{4a^3}S^{TE} \tag{6.4a}$$

$$\alpha_m^{TM} = \frac{m^2\lambda^2}{4a^3}S^{TM}. \tag{6.4b}$$

The transmission of a single mode through these waveguides should have an inverse cube dependence on waveguide height. When more than one mode is present, however, the height dependence is more complex and is given by weighting the losses of each mode. The transmission of a waveguide of length L is given by

$$T = \sum_m f_m \exp(-\alpha_m L)\left(\sum_m f_m\right)^{-1}, \tag{6.5}$$

where α_m is given by equation (6.2) or (6.4) for either the TE or TM mode, and f_m is the fraction of the total input power put into the mth mode.

When the sidewalls are brought close enough together to interact with the incident laser beam, the waveguide can be considered rectangular. As long as b(width of the waveguide) $\gg a$, the total loss α in a rectangular guide can be considered as the sum of the planar waveguide TE loss (α^{TE}) caused by the top and bottom walls and the planar waveguide TM loss (α^{TM}) caused by the sidewalls. Therefore,

$$\alpha = \alpha^{TE} + \alpha^{TM} = \frac{m^2\lambda^2}{4a^3}S^{TE} + \frac{m^2\lambda^2}{4b^3}S^{TM}. \tag{6.6}$$

Bend losses. The ribbon-like design of the waveguide allows flexibility with a bend made in the plane normal to its top and bottom walls, in the geometry shown in figure 6.2. Such a bend preserves the polarization symmetry of the TE-like reflections from the top and bottom walls. In fact, for most practical bend radii, the light propagates in 'whispering gallery mode'. The ray optics picture of this mode of propagation is shown by the arrows in figure 6.2 and corresponds to one in which the light rays reflect from the outer wall only. This is because the bend radius is sufficiently small that the inner wall is bent away from interacting with the light rays. The bend losses can be quantitatively explained using this simple picture.

The additional bend loss α_R can be expressed by $\alpha_B - \alpha_S$, where α_B is the total bend loss and α_S is the straight guide loss. A ray optics analysis shows that α_B can be approximated by $\alpha_B \simeq \alpha_S$ for $R \geq R_0$ and $\alpha_B \simeq S^{TE}/2R$ for $R \leq R_0$, where the cross-over point is given by $R_0 = 2a^3/m^2\lambda^2$. These approximations make it possible to write

$$\alpha_R = \begin{cases} S^{\text{TE}}(R^{-1} - R_0^{-1})/2 & R \le R_0 \\ 0 & R \ge R_0. \end{cases} \tag{6.7}$$

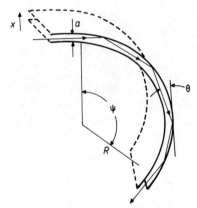

Figure 6.2 The geometry of a bent waveguide, tracing the ray reflections under the conditions of a whispering gallery mode (after Garmire *et al* 1980, © 1980 IEEE).

Twist losses. The ray optics picture of light in a twisted guide shown in figure 6.3 indicates the twist loss. Even if the light is TE polarized in the untwisted guide, the twist introduces some TM polarization into the reflection. The power loss coefficient due to twist, α_T, is given by

$$\alpha_T = a\phi_T^2 S^{\text{TM}}, \tag{6.8}$$

where ϕ_T is the twist rate.

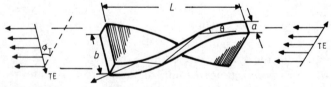

Figure 6.3 The geometry of a waveguide twisted through an angle ϕ_T, showing ray reflections at an angle θ (after Garmire *et al* 1980, © 1980 IEEE).

Besides affecting transmission, a twist causes TE-polarized light at the guide input to rotate in the polarization direction at the output end. This demonstrates that a twisted parallel-plate waveguide can be used as a polarization rotator.

6.1.2.2 Concave-plate waveguides
Concave-plate waveguides were proposed by Nishihara *et al* (1974) for

obtaining a self-focusing effect in transmitted light. Figure 6.4 shows the cross section of the proposed waveguide consisting of two slightly curved parallel plates. The power loss coefficient α_r can be expressed approximately by

$$\alpha_r = \alpha_\infty \left(1 + \frac{3}{\pi}\frac{(\frac{1}{2}a)^{1/2}}{r^{1/2}}\right), \tag{6.9}$$

where a is the distance between plates, r is the curvature of the concave plate, and α_∞ is the power loss coefficient of the parallel plate ($r = \infty$). The second term is negligibly small. For example, when $a = 0.3$ mm, $r = 1$ m, the contribution of the second term is 1%.

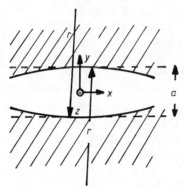

Figure 6.4 The cross section of a waveguide consisting of two slightly curved parallel plates.

The focusing effect of this waveguide can be expressed using the beam width w. An approximate relation can be obtained by analyzing the modes in an extremely flat elliptical waveguide. The beam width w is calculated as

$$w = (2\sqrt{2}/\sqrt{\pi})(a/2)^{3/4}r^{1/4} \tag{6.10}$$

for the TE_1 fundamental mode. A numerical example is $w = 2.4$ mm when $a = 0.3$ mm and $r = 1$ m. It can be seen that even a slight curvature has a self-focusing effect. This has been demonstrated experimentally by Nishihara *et al* (1974).

6.1.2.3 Fabrication and transmission properties

High power laser light was first transmitted by Garmire *et al* (1979) through the parallel-plate metallic waveguide shown in figure 6.1. In their experiment, the output of a 250 W CO_2 laser was focused by a cylindrical ZnSe lens into the waveguide. More than 200 W c.w. was transmitted through the waveguide, representing an 80% overall efficiency. This power level was reported to be limited solely by the output of the CO_2 laser used. Noticeable heating of the guide walls occurred only at the entrance where coupling was not optimized. Therefore, Garmire *et al* (1980) concluded that kilowatts of c.w. power at 10.6 μm can be transmitted.

Matsushima *et al* (1981) have proposed plastic waveguides coated with evaporated aluminum. The basic idea of this technique is to employ plastic plates with sufficient flexibility on which a highly reflective metal is deposited by evaporation. The waveguide is thus more flexible than the metallic waveguide. Two types of waveguide, planar and rectangular, have been constructed and tested. The measured transmittance for the planar waveguide with 1 mm thickness and 11 mm width was 85%, while the calculated value is 99.9%. The difference is considered to be mostly due to local thickness variation. 78% and 90% transmittance through a 37 cm long waveguide were obtained under the bend of 25 cm and 50 cm radius of curvature, respectively. The values are normalized to the straight guide transmittance of 85%. The transmittance of the rectangular waveguide was nearly the same as that of the planar waveguide.

Recently, parallel-plate metallic waveguides were extensively studied by Kubo and Hashishin (1986). Figure 6.5 shows the fabricated waveguide 1 m long. The measured transmission coefficient was 85%, and a

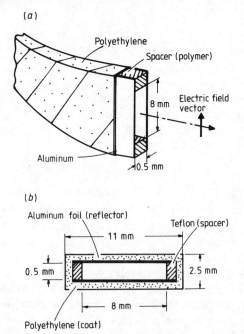

Figure 6.5 The structure of a parallel-plate metallic waveguide developed by Kubo *et al*: (*a*) overview; (*b*) cross section (after Kubo and Hashishin 1986).

maximum laser power of 95 W was transmitted. Furthermore, the output laser beam was focused by the combination of the cylindrical lens and a conventional lens resulting in a beam diameter of 0.5 mm and a power density of 40 kW cm^{-2}.

Besides power transmission, Mizushima *et al* (1980) discussed the possibility of long haul transmission using the parallel-plate waveguide. If the waveguide is used for undersea optical communication, the radius of curvature of the waveguide is expected to be more than 1000 m, resulting in ultra-low transmission loss. The other advantage is its single-mode nature. The wavelength-dependent part of the propagation velocity is very small; the contribution is less than 10^{-8} for $\Delta\lambda/\lambda = 0.1$, so that the propagation delay difference is negligible. Difference between modes is also small; it originates from the very small skin thickness. Experimental studies, however, have not yet been reported.

6.1.3 Circular waveguides
6.1.3.1 Basic structure
Circular hollow metallic waveguides were first studied in the 1950s, mainly for optical communications (Karbowiak 1958). Later Marcatili and Schmeltzer (1964) discussed the possibility of light transmission through metallic and dielectric hollow waveguides. The loss was estimated to be 1.8 dB km^{-1} when the TE$_{01}$ fundamental mode light at 1 μm wavelength was transmitted through a copper pipe with 0.5 mm diameter. This loss value seems relatively low. However, the study of these waveguides became inactive because of the appearance of glass optical fibers whose transmission losses were expected to be even lower. Instead, these hollow waveguides are being studied from the viewpoint of application for infrared transmission.

In this section the formulation developed by Garmire (1976) is described. In a straight hollow cylindrical waveguide with sidewalls of complex refractive index $\tilde{n}$ $(= n - i\kappa)$, the power loss coefficient for guided light is

$$\alpha_{pq} = 2\left(\frac{u_{pq}}{2\pi}\right)^2 (Re\,\tilde{n}_p)\frac{\lambda^2}{a^3}, \qquad (6.11a)$$

where

$$\tilde{n}_p = \begin{cases} (\tilde{n}^2 - 1)^{-1/2} & \text{TE}_{0q}\text{ modes} \\ \tilde{n}^2(\tilde{n}^2 - 1)^{-1/2} & \text{TM}_{0q}\text{ modes} \\ \frac{1}{2}(\tilde{n}^2 + 1)(\tilde{n}^2 - 1)^{-1/2} & \text{EH}_{pq}\text{ modes, } p \neq 0, \end{cases} \qquad (6.11b)$$

a is the radius of the hollow part, λ the wavelength, and u_{pq} the geometrical mode factor, the qth root of the Bessel function $J_{p-1}(u_{pq}) = 0$.

Equation (6.11) has a range of validity given by

$$a/\lambda \gg |\tilde{n}|\frac{u_{pq}}{2\pi}. \tag{6.12}$$

This condition is satisfied only for highly multimode guides since numerically it requires $a \gg 40\lambda$ for the lowest mode in copper.

For waveguides with metal walls guiding light in the infrared near 10.6 μm, $|n^2 - \kappa^2| \gg 1$, so that $(n^2 - 1)^{1/2} \simeq \tilde{n}$, and

$$Re\ \tilde{n}_p = \begin{cases} n(n^2 + \kappa^2)^{-1} & \text{TE modes} \\ n & \text{TM modes} \\ \frac{1}{2}n & \text{EH modes.} \end{cases} \tag{6.13}$$

Since n, $\kappa \gg 1$, only the TE modes have a small value of $Re\ \tilde{n}_p$ and correspondingly low loss. The power loss coefficient in a copper waveguide ($\tilde{n} = 13 - i65$) for the lowest order TE mode at 10.6 μm is

$$\alpha_{01}^{\text{TE}} = 0.0024\ \lambda^2/a^3. \tag{6.14}$$

From this expression it is expected that a waveguide of 200 μm radius will transmit 97% of the intensity of this mode over one meter.

These results indicate that the TE_{0q} modes of a circular hollow waveguide can be used for high power laser light transmission. However, low loss TE_{0q} modes in circular metallic waveguides have a donut-shaped intensity profile and are azimuthally polarized at every point, as shown in figure 6.6. Such distributions of optical electric fields cannot be excited efficiently with the output of commercially available lasers.

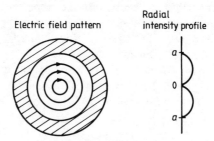

Electric field pattern   Radial intensity profile

Figure 6.6 The TE_{01} mode in hollow, cylindrical, metallic waveguides (after Garmire *et al* 1976a).

Furthermore, it has been pointed out by Marhic (1981) and Miyagi *et al* (1984) that the bending loss caused by the mode coupling is considerably high because of the near degeneracy of the TE and TM modes. For these reasons there still exists some difficulty in applying these waveguides to infrared light transmission.

An experimental result using hollow circular metal waveguides was reported by Nakatsuka and Kubo (1979). In their experiment, a commercially available copper pipe of diameter 1.2 mm was used. The CO_2 laser light was focused into the pipe using a Ge lens. It was found that the smoothness of the inner surface of the pipe is particularly important in reducing the loss. 20% of the input laser power (the maximum power was 20 W) was transmitted through the pipe, which was 0.8 m long.

6.1.3.2 Helical-circular waveguides

Marhic *et al* (1978a,b, Marhic 1979) and Casperson and Garfield (1979) proposed a waveguide based on the 'whispering gallery' principle, which is described in section 6.1.2.1 in connection with the discussion about the bending loss of the parallel-plate waveguide. The advantage of this helical-circular waveguide is that the transmission loss is expected to be low because the grazing reflectivity in the whispering gallery configuration is quite high.

A light ray in a helical-circular waveguide runs along the surface of the hollow waveguide as shown in figure 6.7. In the figure the solid line shows the light ray which reflects at a grazing angle along the dotted line on the inner surface of the tube. According to Marhic (1979), there are two orthogonal eigenpolarization states, and the power loss coefficients for these states can be approximated by

$$\alpha_+ \simeq 4\rho^{-1} Re[\tilde{n}^2(\tilde{n}^2 - 1)^{-1/2}] = \alpha^{TM}$$

$$\alpha_- \simeq (1 + \sigma^2) 4\rho^{-1} Re[(\tilde{n}^2 - 1)^{-1/2}] = (1 + \sigma^2)\alpha^{TE},$$

(6.15)

where α^{TM} and α^{TE} mean the power loss coefficients for TM and TE modes of two-dimensional trajectories with the same curvatures, which correspond to light rays travelling a helical tube of zero pitch. $\tilde{n}$ is the complex refractive index, and σ is defined as (ρ: radius of curvature of the helix) $\times$ (τ: torsion). In figure 6.7, $\rho = (a^2 + b^2)/a$, and $\tau = b/(a^2 + b^2)$.

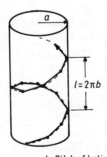

l : Pitch of helix

Figure 6.7 The helical-circular waveguide. The solid line shows the light ray.

It should be noted that equation (6.15) is valid only for infrared and far-infrared radiation propagating inside the metallic guides. Equation (6.15) shows that the power loss coefficient for the $(-)$ eigenpolarization that is almost TE mode is about $(1 + \sigma^2)$ times higher than that of the low loss TE waves. This is important since it shows that the minimum losses of metallic waveguides for the infrared and far-infrared regions of the spectrum will increase drastically as σ approaches and exceeds unity (for a fixed ρ), and this fact will have to be taken into account when designing waveguides of this type.

Various waveguides using the whispering gallery principle have been proposed to date. Figure 6.8 shows possible waveguide configurations (Casperson and Garfield 1979).

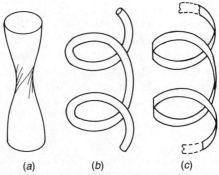

(a)   (b)   (c)

Figure 6.8 Possible waveguide configurations for whispering-gallery operation: (a) twisted metal tube of elliptical cross section; (b) helical metal tube of circular cross section; (c) helical metal strip (after Casperson and Garfield 1979, © 1979 IEEE).

6.1.4 Dielectric-coated metallic waveguides

6.1.4.1 Design theory

The various hollow metallic waveguides described so far cannot transmit the hybrid HE_{11} mode with low loss. Therefore, it is difficult to launch the light from commercially available CO_2 lasers whose light field distribution is similar to the HE_{11} mode. Miyagi and Kawakami (1981, 1984) and Miyagi et al (1983b) proposed and developed dielectric-coated circular metallic waveguides whose HE_{11} transmission losses are expected to be low. The design theory for the waveguides is described here (Miyagi and Nishida 1986).

The structure of the dielectric-coated circular metallic waveguide is shown in figure 6.9. Two kinds of dielectrics with refractive indices of $\eta_i n_0$ ($i = 1, 2$) and a thickness of $\delta_i a$ ($i = 1, 2$) are coated alternately. The hollow region has a radius of a and a refractive index of $n/n_0 = 1$.

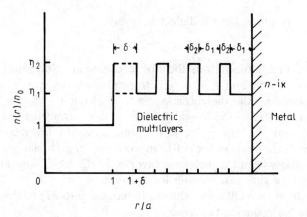

Figure 6.9 The structure of a dielectric-coated circular metallic waveguide (after Miyagi and Kawakami 1984, © 1984 IEEE).

The metallic wall has a complex refractive index $n - i\kappa = n_0(n' - i\kappa')$. The refractive index of the layer adjacent to the metal is $\eta_i n_0$ and $\eta_1 < \eta_2$, and the thickness of the layer adjacent to the air is δa. Here, the thickness of the dielectric layers must satisfy

$$\delta_i(\eta_i^2 - 1)^{1/2} n_0 k_0 a = \frac{\pi}{2} \quad (i = 1, 2), \quad (6.16)$$

where k_0 is the wavenumber, so as to act as the impedance or admittance transformer in the transmission line model (Miyagi *et al* 1983b). When the total number m of dielectric layers is assumed to be odd:

$$m = 2m_p + 1, \quad (6.17)$$

transmission losses of the hybrid HE modes are minimized under the condition that the thickness δa satisfies

$$\delta(\eta_1^2 - 1)^{1/2} n_0 k_0 a = \pm \tan^{-1}\left[\frac{\eta_1}{(\eta_1^2 - 1)^{1/4}}\left(\frac{\eta_1}{\eta_2}\right)^{m_p} C^{-m_p/2}\right] + s\pi. \quad (6.18)$$

Here, s is an integer and C is defined by

$$C = (\eta_1^2 - 1)/(\eta_2^2 - 1). \quad (6.19)$$

The power loss coefficient α_∞ of the HE$_{11}$ mode can be expressed by

$$\alpha_\infty = n_0 k_0 \frac{u_0^2}{(n_0 k_0 a)^3} F_{\min}, \quad (6.20)$$

where u_0 is 2.405 and $F_{\min}$ is

$$F_{\min} = \frac{1}{2} \frac{n}{n^2 + \kappa^2} C^{m_p}\left[1 + \frac{\eta_1^2}{(\eta_1^2 - 1)^{1/2}}\left(\frac{\eta_1}{\eta_2}\right)^{2m_p} C^{-m_p}\right]^2. \quad (6.21)$$

When the waveguide has no dielectric layers, F_{min} is

$$F_{min} = \tfrac{1}{2}n. \qquad (6.22)$$

It can be understood from the above discussion that since $\kappa \gg n$ in the infrared region and $C < 1$, the transmission loss can be drastically reduced by adding the dielectric layers. For example, when $a = 500\ \mu m$ and $n - i\kappa = 20.5 - i58.6$ (aluminum), the transmission loss is $0.08\ dB\ m^{-1}$ for the waveguide having one dielectric layer composed of germanium (refractive index 4.0). In contrast, the transmission loss of the waveguide without a dielectric layer is $12\ dB\ m^{-1}$. Note also that in order to reduce the transmission loss further, $n/(n^2 + \kappa^2) \equiv F_{metal}$ must be as small as possible, as shown in equation (6.21). Silver is most appropriate for reducing the loss.

Further, it should be noted that when the following condition is satisfied for the dielectric layer ($\delta^* a$ is the thickness and ηn_0 is the refractive index)

$$\delta^*(\eta^2 - 1)^{1/2}n_0 k_0 a = s\pi \qquad (s \text{ is an integer}) \qquad (6.23)$$

the presence of the material can be completely ignored. Even if the material has a large absorption loss, the transmission loss of the waveguide remains small.

6.1.4.2 Fabrication and transmission properties

Miyagi *et al* (1983a) developed a method for the fabrication of dielectric-coated metallic waveguides. The fabricated waveguide consists of a nickel metal pipe with a germanium coating. The method is based on RF sputtering, plating, and etching techniques, as shown schematically in figure 6.10. Germanium is sputtered onto a polished aluminum pipe (outer diameter 1.5 mm, thickness 0.2 mm, and length 1.2 m) which is rotated at 10 rpm and moved axially at 10 cm min^{-1} in a vacuum chamber (2.0×10^{-2} Torr of Ar gas) to obtain a uniform thin layer. A schematic view of the sputtering apparatus is shown in figure 6.11. It can be seen by X-ray analysis that the sputtered germanium on the aluminum pipe is amorphous. The sputtering rate is around $0.15\ \mu m\ h^{-1}$ for 500 W RF power. The typical thickness of the germanium layer is designed to be $0.45\ \mu m$, so as to satisfy the minimum loss condition.

In order to form a metallic pipe, nickel is plated directly onto the germanium layer. A nickel plate anode is formed into a circular cylinder of diameter 7.5 cm and length 1.2 m, and the cathode for plating is located at its center so that the nickel is deposited uniformly. The plating bath is composed of $NiSO_46H_2O$ (240 g), $NiCl_26H_2O$ (45 g) and H_3BO_4 (30 g) in 1000 cm^3 water. A nickel layer of 70–200 μm is formed with sufficient mechanical strength. Finally, the aluminum pipe is etched away by a 10–20% NaOH solution and a circular hollow nickel waveguide with an inner germanium layer is fabricated.

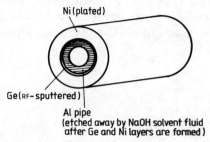

Figure 6.10 The method used for fabricating germanium-coated nickel waveguides (after Miyagi *et al* 1983a).

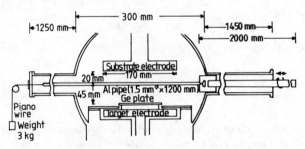

Figure 6.11 A schematic view of the sputtering apparatus for forming a germanium layer (after Miyagi *et al* 1983a).

The total transmission losses of fabricated waveguides are always below $0.5 \, \mathrm{dB \, m^{-1}}$ for a 1 m long waveguide at 10.6 μm wavelength. This loss includes the launching loss. In contrast, the loss of the waveguide with no dielectric layers is around $2.5 \, \mathrm{dB \, m^{-1}}$. Therefore, it can be seen that the losses of metallic waveguides can be reduced significantly by coating with a dielectric material.

However, theory predicts that the power loss of the HE_{11} mode is $3.2 \times 10^{-2} \, \mathrm{dB \, m^{-1}}$. The difference between the theoretical and experimental results may be caused by the roughness or gradual deformation of the waveguides and deviation from ideal of the thickness of the germanium layer.

Bending losses measured in nickel waveguides without and with an inner germanium layer are shown in figure 6.12, where the first 20 cm is made straight and the other 80 cm is bent with a uniform curvature. When the polarization is parallel to the plane of curvature (denoted by

$E_\parallel$), transmission is extremely small for the nickel waveguide, even with a large bending radius, whereas it is considerably greater for the germanium-coated waveguide, as well as when the polarization is perpendicular. Miyagi *et al* (1983a) insist that if dielectric-coated waveguides with smaller losses can be realized in straight structures, the waveguides can be much more sharply bent without significant loss increase.

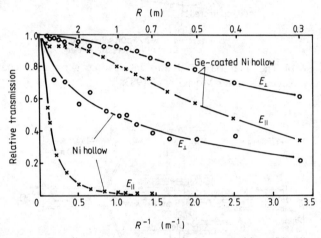

Figure 6.12 Bending losses of a nickel waveguide with diameter 1.5 mm and length 1.02 m whose straight-waveguide loss is 2.85 dB, and those of a germanium-coated nickel waveguide with diameter 1.5 mm and length 1.01 m whose straight-waveguide loss is 0.43 dB. $E_\parallel$ and $E_\perp$ denote the electric fields whose polarization vector is parallel and perpendicular to the plane of curvature, respectively (after Miyagi *et al* 1983a).

6.1.5 Summary

Metallic hollow waveguides can be divided into two groups: one is parallel-plate waveguides and the other is circular hollow waveguides. Among the parallel-plate waveguides, various structures such as the concave-plate waveguide (Nishihara *et al* 1974) have been proposed. More than 200 W c.w. CO_2 laser power has been transmitted through the parallel-plate waveguide, representing 80% overall transmittance (Garmire *et al* 1979). In addition, Kubo and Hashishin (1986) succeeded in constructing a machine for laser surgery with a parallel-plate metallic waveguide.

Circular metallic waveguides have a low loss for TE mode operation, whereas the hybrid HE modes which can be excited efficiently by commercially available lasers give a high transmission loss. Miyagi and colleagues (1981, 1983a and 1984a,b) developed a dielectric-coated circular hollow metallic waveguide whose HE mode loss can be significantly lowered. A nickel pipe waveguide with a germanium inner coating gave a low loss of $0.35 \, \mathrm{dB \, m}^{-1}$ at $10.6 \, \mu$m wavelength.

Also, helical-circular waveguides have been proposed which are based on the whispering gallery principle, i.e., grazing reflections only at one-sided metallic surfaces. This waveguide is basically expected to give low loss and flexibility. However, experiments on the transmission properties using a CO_2 laser have not yet been reported.

Hollow metallic waveguides have great potential for transmitting high power laser light.

6.2 Dielectric hollow waveguides

6.2.1 Introduction
In this section, hollow waveguides composed of various dielectric materials are presented. They are mainly studied for laser power transmission. The first dielectric hollow waveguides studied were the ones whose transmission mechanism is the same as metallic hollow waveguides described in the previous section. Then dielectric leaky waveguides were proposed by Miyagi (1981) and Miyagi and Nishida (1980a,b) in order to obtain a lower loss than that of the dielectric hollow waveguides.

Hidaka *et al* (1982, Hidaka 1982) proposed hollow optical fibers whose transmission mechanism is largely different from that of the other hollow waveguides. These fibers can transmit infrared light by total reflection between the hollow core and the dielectric glass cladding. This guiding mechanism is possible only in the wavelength region of abnormal dispersion where the refractive index is lower than unity.

Liquid-core fibers have also been proposed for infrared light transmission (Majumdar *et al* 1979).

6.2.2 Dielectric hollow grazing waveguides
The light guiding mechanism of the circular hollow waveguides composed of dielectric materials is based on the grazing reflection between the dielectric cladding and the hollow core, which is the same as in the metallic hollow waveguides described in section 6.1.3. The transmission losses of the dielectric hollow waveguides increase considerably as the bending radius decreases, and so they are not readily applicable to flexible infrared light transmission. The theoretical loss estimation will be described in the following section in connection with the estimation in dielectric leaky waveguides.

Only a few reports on the experimental results of dielectric hollow grazing waveguides have been presented to date. Jenkins and Devereux (1986) gave the experimental results of a curved hollow silica waveguide using a CO_2 laser and compared them with theory. Bornstein *et al* (1985) and Bornstein and Croitoru (1986) reported the use of chalcogenide hollow fibers for infrared transmission. They discussed their, experimental results on the basis of dielectric leaky waveguides.

6.2.3 Dielectric leaky waveguides

Miyagi and Nishida (1980a) proposed dielectric leaky waveguides in order to improve transmission properties of dielectric hollow waveguides. In this section their transmission properties are described in connection with those of the dielectric hollow grazing waveguides.

The geometry of the dielectric leaky waveguide proposed by Miyagi and Nishida (1980b) is shown in figure 6.13. The radius of the hollow part is a, and the thickness of the dielectric tube is δa. The refractive index of the dielectric tube is ηn_0 and of the other part is n_0. This leaky waveguide is characterized by its thin tube layer, while the dielectric hollow grazing waveguides have thick dielectric layers whose outer surfaces never affect the transmission properties. In this structure, there exist bound or surface modes which are trapped in the region $a < r < (1 + \delta)a$. However, these modes are damped out when the material has considerable loss. In contrast, the modes whose energy is trapped in $r < a$ are expected to be low. These modes are called 'leaky modes'.

Using a transverse transmission-line model, we can obtain the minimum power loss coefficient which is attainable in each mode (Miyagi and Nishida 1980b). That is,

$$\alpha_{\min} = 2\frac{n_0 k_0}{\eta^2 - 1}\frac{u_\infty^3}{(n_0 k_0 a)^4}$$

$$\times \begin{cases} 1 & \text{TE}_{0m} \text{ mode} \\ \eta^4 & \text{TM}_{0m} \text{ mode} \\ \frac{1}{2}(\eta^4 + 1) & \text{HE}_{nm} \text{ and EH}_{nm} \text{ modes,} \end{cases} \tag{6.24}$$

where k_0 is the wavenumber of the vacuum, and u_∞ is the root of the Bessel function and is approximated by

$$u_\infty = \begin{cases} (2m + \frac{1}{2})\dfrac{\pi}{2} & \text{TE}_{0m} \text{ and TM}_{0m} \text{ modes} \\[2mm] (n + 2m - \frac{3}{2})\dfrac{\pi}{2} & \text{HE}_{nm} \text{ mode} \\[2mm] (n + 2m + \frac{1}{2})\dfrac{\pi}{2} & \text{EH}_{nm} \text{ mode.} \end{cases} \tag{6.25}$$

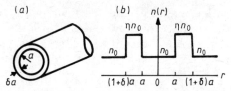

Figure 6.13 (*a*) The geometry of a dielectric tube waveguide, and (*b*) its refractive index profile, where $\eta > 1$ (after Miyagi *et al* 1980b, © 1980 IEEE).

As shown, the losses can be made arbitrarily small by choosing the radius a of the tube to be sufficiently large relative to the wavelength λ. Furthermore, it can be shown that the mode with the lowest attenuation is the HE_{11} mode (almost linearly polarized) if $\eta < 1.63$, and the TE_{01} mode (circularly polarized) if $\eta > 1.63$. In figure 6.14, the power loss coefficient of the lowest order HE_{11} mode is shown as a function of the core radius a for $\eta = 1.5$ and $n_0 = 1$. For example, in order to design a waveguide with loss coefficient of 5 dB km^{-1}, $a = 600$ μm for $\lambda = 10$ μm is needed. These values can be compared with those of a dielectric hollow grazing waveguide whose configuration is shown in figure 6.15. The power loss coefficient α_H of the dielectric hollow grazing waveguide is described by

$$\alpha_H = 2\frac{n_0 k_0}{(\eta^2 - 1)^{1/2}} \frac{u_\infty^2}{(n_0 k_0 a)^3}$$

$$\times \begin{cases} 1 & \text{TE}_{0m} \text{ mode} \\ \eta^2 & \text{TM}_{0m} \text{ mode} \\ \tfrac{1}{2}(\eta^2 + 1) & \text{HE}_{nm} \text{ and EH}_{nm} \text{ modes.} \end{cases} \quad (6.26)$$

Therefore, the ratio $\alpha_{\min}/\alpha_H$ of the power loss coefficients of the corresponding modes in the two waveguides is

$$\frac{\alpha_{\min}}{\alpha_H} = \frac{1}{(\eta^2 - 1)^{1/2}} \frac{u_\infty}{n_0 k_0 a}$$

$$\times \begin{cases} 1 & \text{TE}_{0m} \text{ mode} \\ \eta^2 & \text{TM}_{0m} \text{ mode} \\ \dfrac{\eta^4 + 1}{\eta^2 + 1} & \text{HE}_{nm} \text{ and EH}_{nm} \text{ modes.} \end{cases} \quad (6.27)$$

One finds that the power loss coefficient of the dielectric leaky waveguide is smaller than that of the dielectric hollow grazing waveguide by several orders of magnitude, because a/λ is of the order of several tens or hundreds in practical guide parameters. Furthermore, the suppression for the higher order modes is greater than in the dielectric hollow grazing waveguide because the power loss coefficient is approximately proportional to $(n + 2m)^3$, not to $(n + 2m)^2$, as seen in equations (6.24) and (6.25).

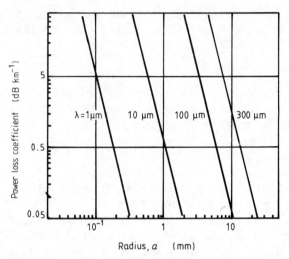

Figure 6.14 Attenuation of the HE$_{11}$ mode as a function of core radius a for $\eta = 1.5$ and $n_0 = 1$ (after Miyagi and Nishida 1980b, © 1980 IEEE).

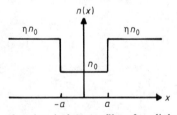

Figure 6.15 The refractive index profile of a dielectric hollow grazing waveguide.

The bending loss has been estimated by Miyagi (1981). It was shown that the dielectric leaky waveguide can be bent with a smaller bending radius than the metallic hollow and dielectric hollow grazing waveguides when the guide material is appropriately selected.

It can be concluded from the above discussion that the dielectric leaky waveguide can transmit the hybrid HE$_{11}$ mode with low loss. This is preferable for use with lasers because commercially available lasers have the same light field patterns as the hybrid HE$_{11}$ mode light.

6.2.4 Hollow-core optical fibers

Hidaka (1982) and Hidaka *et al* (1981, 1982) proposed a new type of infrared optical waveguide made with oxide glass. The principle is shown in figure 6.16. When the refractive index n of the cladding material is smaller than unity, the obliquely incident light, whose incident angle from air to the cladding surface is larger than the critical angle $\theta_T = \sin^{-1}n$, is totally reflected. Thus, it would be expected that hollow-core optical fibers with such cladding would show low transmission loss at the wavelength with $n < 1$. However, it should be noted that the refractive index smaller than unity at the middle-infrared region is originated from the strong absorption due to lattice vibrations. This situation is shown in figure 6.17. n is given by

$$n(\omega) = 1 + \frac{c}{\pi}\int_0^\infty \frac{\omega_i \kappa(\omega_i)}{\omega_i^2 - \omega^2}\mathrm{d}\omega_i, \tag{6.28}$$

where c is the velocity of light, and κ is the imaginary part of the

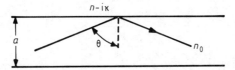

Figure 6.16 Hollow-core, oxide-glass-cladding optical fiber. The refractive index of the core n_0 is 1.0. When the refractive index of the cladding $n < 1.0$, waveguide transmission loss may be small.

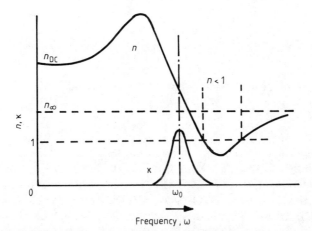

Figure 6.17 The real part n and imaginary part κ of the refractive index $\tilde{n}$. n smaller than unity is due to large κ. n_{DC} and n_∞ are the refractive indices of light with frequency $= 0$ and ∞ (after Hidaka 1982).

complex refractive index $\tilde{n} = n - i\kappa$. κ is proportional to the absorption coefficient α through $\kappa = (c/2\omega)\alpha$. For the $n < 1$ middle-infrared wavelength region, κ inevitably shows a large value. Thus we cannot obtain perfect total reflection at the boundary of such hollow-core optical fibers, since the large κ value leads to large waveguide transmission loss.

The transmission loss of this fiber can be estimated by mode analysis (Hidaka 1982), as shown in figure 6.18. The solid lines show the HE_{11} mode losses, and the dashed line the maximum allowable loss for practical application. As shown, the loss increases rapidly as n approaches unity from below.

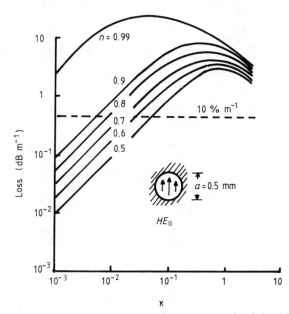

Figure 6.18 Loss characteristics of the HE_{11} mode (after Hidaka 1982).

It is important to select the appropriate glass material in order to obtain low loss at the CO_2 laser wavelength. Figure 6.19 shows the real part n and imaginary part κ of the complex refractive index of fused SiO_2 and those of pure GeO_2 glass (Hidaka *et al* 1982). n and κ are calculated from the observed reflectance. The frequency characteristics of n and κ of pure GeO_2 glass are very similar to those of pure fused SiO_2. The peak position of GeO_2 is shifted by 200 cm^{-1} to a wavenumber lower than that of SiO_2. The minimum value of n of GeO_2 is about

0.2 at 935 cm^{-1}, whereas the maximum of κ is about 1.7 at 830 cm^{-1}. The absorption profile is somewhat sharper in GeO$_2$ than in SiO$_2$; hence the minimum n of GeO$_2$ is smaller than that of SiO$_2$ from equation (6.28). The transmission loss of a waveguide made from pure GeO$_2$ will exhibit its maximum value at 980 cm^{-1} (10.2 μm) due to the frequency characteristics of n and κ, which is slightly higher than the 940 cm^{-1} (10.6 μm) laser wavenumber. However, it is possible to shift the wavenumber to give the minimum loss by doping with several metal oxides, such as ZnO and K$_2$O.

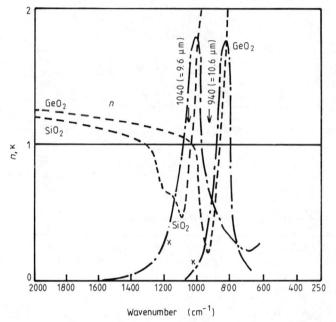

Figure 6.19 n and κ of pure SiO$_2$ and GeO$_2$ glasses. Values of n are smaller than unity near the CO$_2$ laser wavenumber (940 cm^{-1}), but the wavenumber giving the minimum waveguide transmission loss is not coincident with the laser wavenumber (after Hidaka *et al* 1982).

Among the several glass compositions, GeO$_2$–ZnO–K$_2$O glass is suitable for infrared transmitting hollow-core fibers, since the dopant K$_2$O can shift a minimum loss position toward small wavenumbers and the dopant ZnO acts as a glass stabilizer. Figure 6.20 shows the estimated minimum transmission loss (dashed line) and its corresponding wavenumber (solid line). The hollow waveguide bore size is assumed to be 1 mm, and the transmission mode is HE$_{11}$. As shown, the wavenumber giving the minimum loss covers the range 850–980 cm^{-1}, which means that the 940 cm^{-1} CO$_2$ laser wavenumber can be obtained. Transmission loss along with the 940 cm^{-1} line changes from 0.05 dB m^{-1} (K$_2$O rich) to 0.15 dB m^{-1} (ZnO rich); therefore the K$_2$O rich glass is preferable.

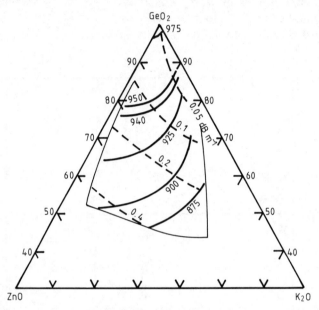

Figure 6.20 Minimum waveguide transmission loss of 1 mm diameter hollow fiber. The dashed lines are the loss values and the solid lines are their wavenumbers (after Hidaka *et al* 1982).

CO$_2$ laser light transmission experiments have been performed by using an 80 mol.% GeO$_2$–10 mol.% ZnO–10 mol.% K$_2$O glass. In this composition, the expected transmission loss at 940 cm^{-1} (10.6 μm) is about 0.1 dB m^{-1} for the 1 mm bore waveguide. Spectroscopic n and κ data are about 0.6 and 0.1, respectively, at 940 cm^{-1}. Figure 6.21 shows the expected (theoretical) loss versus wavenumber. The loss at a wavenumber above 950 cm^{-1} increases by a factor of 10 compared to the minimum loss at 950 cm^{-1} due to the increase of n with wavenumber. The strong wavenumber dependence of the loss is one of the most distinctive features of these optical fibers. The experimental results are also shown as solid points. The wavenumber dependence of the loss coincides with the theoretical value, although the experimental loss at 940 cm^{-1} is about 20 times larger than the theoretical one. One of the reasons for this may be imperfections in the inner wall of the waveguide used. The transmitted mode can be estimated to be the HE$_{11}$ mode from the sharp output pattern.

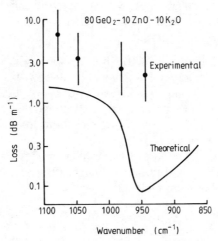

Figure 6.21 The expected (calculated) wavenumber characteristics (solid line) of the loss of the 1 mm diameter hollow fiber. The mode is HE$_{11}$ type. The solid points show the experimental data (after Hidaka *et al* 1982).

Hidaka *et al* (1982) concluded that if a smooth inner surface is obtained, the transmission loss can be lowered further, say $0.1 \, \text{dB m}^{-1}$. Characterization of such hollow-core optical fibers has also been performed by Scheggi *et al* (1985). Furthermore, it should be noted that a vacuum ultraviolet light transmission using hollow-core quartz-glass cladding optical fibers has been reported by Watanabe *et al* (1984). In this case, the refractive index dispersion due to the electronic transition is used instead of the lattice vibration in the infrared optical fibers.

6.2.5 Other hollow optical fibers
Some hollow fibers have their core filled with liquid. The advantages of liquid-core fibers are: suitable low loss organic liquids can be chosen; there are no stress effects leading to birefringence; and wall imperfections and scattering effects are negligible. A tetrachloroethylene (C_2Cl_4)-filled fused-quartz fiber has been reported by Majumdar *et al* (1979). Although the transmission loss seems high (Takahashi *et al* 1985), these fibers have potential for transmitting infrared light.

6.2.6 Summary
Dielectric hollow waveguides are divided into the following classes:

(i) dielectric hollow waveguides which can transmit the light by grazing reflection;

(ii) dielectric leaky waveguides;

(iii) hollow-core optical fibers which can transmit the light by total reflection in the abnormal dispersion region;

(iv) liquid-core optical fibers.

The advantage of these dielectric hollow waveguides is that they are relatively flexible compared to metallic hollow waveguides. Of the dielectric hollow waveguides, dielectric leaky waveguides and hollow-core optical fibers are particularly advantageous because they can transmit the low loss HE mode whose joint loss with commercially available CO_2 lasers is low.

The minimum transmission loss obtained by experiment is $20 \, \mathrm{dB \, m^{-1}}$ for the hollow-core optical fibers, although the predicted loss is less than $0.1 \, \mathrm{dB \, m^{-1}}$ at a $10.6 \, \mu\mathrm{m}$ wavelength.

References

Bornstein A and Croitoru N 1986 *Appl. Opt.* **25** 355–8

Bornstein A, Croitoru N and Seidman 1985 *Appl. Phys. Lett.* **46** 705–7

Casperson L W and Garfield T S 1979 *IEEE J. Quantum Electron.* **QE-15** 491–6

Garmire E 1976 *Appl. Opt.* **15** 3037–9

Garmire E, McMahon T and Bass M 1976a *Appl. Opt.* **15** 145–50

—— 1976b *Appl. Phys. Lett.* **29** 254–6

—— 1979 *Appl. Phys. Lett.* **34** 35–7

—— 1980 *IEEE J. Quantum Electron.* **QE-16** 23–32

Hidaka T 1982 *J. Appl. Phys.* **53** 93–7

Hidaka T, Kumada K, Shimada J and Morikawa T 1982 *J. Appl. Phys.* **53** 5484–90

Hidaka T, Morikawa T and Shimada J 1981 *J. Appl. Phys.* **52** 4467–71

Jenkins R M and Devereux W J 1986 *IEEE J. Quantum Electron.* **QE-22** 718–22

Karbowiak A E 1958 *Proc. IRE* **41** 1706–11

Kubo U and Hashishin Y 1986 *Technical Digest of 22nd Symp. in Institute of Electrical Communication, Tohoku University, Sendai, Japan* 43–52 (in Japanese)

Majumdar A K, Hinkley E D and Menzies R T 1979 *IEEE J. Quantum Electron.* **QE-15** 408–10

Marcatili E A J and Schmeltzer R A 1964 *Bell Syst. Tech. J.* **43** 1783–809

Marhic M E 1979 *J. Opt. Soc. Am.* **69** 1218–26

—— 1981 *Appl. Opt.* **20** 3436–41

Marhic M E and Garmire E 1981 *Appl. Phys. Lett.* **38** 743–5

Marhic M E, Kwan L I and Epstein M 1978a *Appl. Phys. Lett.* **33** 609–11

—— 1978b *Appl. Phys. Lett.* **33** 874–6

Matsushima T, Yamauchi I and Sueta T 1981 *Japan. J. Appl. Phys.* **20** 1345–6

Miyagi M 1981 *Appl. Opt.* **20** 1221–9

Miyagi M, Harada K and Kawakami S 1984 *IEEE Trans. Microwave Theory and Tech.* **MTT-32** 513–21

Miyagi M, Hongo A, Aizawa Y and Kawakami S 1983a *Appl. Phys. Lett.* **43** 430–2

Miyagi M, Hongo A and Kawakami S 1983b *IEEE J. Quantum Electron.* **QE-19** 136–45

Miyagi M and Kawakami S 1981 *Appl. Opt.* **20** 4221–6

—— 1984 *J. Lightwave Technol.* **LT-2** 116–26

Miyagi M and Nishida S 1980a *IEEE Trans. Microwave Theory Tech.* **MTT-28** 398–401

—— 1980b *IEEE Trans. Microwave Theory Tech.* **MTT-28** 536–41

—— 1986 *Technical Digest of 22nd Symp. in Institute of Electrical Communication, Tohoku University, Sendai, Japan* 53–61 (in Japanese)

Mizushima Y, Sugeta T, Urisu T, Nishihara H and Koyama J 1980 *Appl. Opt.* **19** 3259–60

Nakatsuka M and Kubo U 1979 *Technical Digest of Annual Meeting of Institute of Electrical Engineering of Japan* 392 (in Japanese)

Nishihara H, Inoue T and Koyama J 1974 *Appl. Phys. Lett.* **25** 391–3

Scheggi A M, Falciai R and Gironi G 1985 *Appl. Opt.* **24** 4392–4

Takahashi H, Sugimoto I, Takabayashi T and Yoshida S 1985 *Opt. Commun.* **53** 164–8

Watanabe M, Hidaka T, Tanino H, Hoh K and Mitsuhashi Y 1984 *Appl. Phys. Lett.* **45** 725–7

7 Applications of Infrared Optical Fibers

In this chapter, various applications of infrared optical fibers are described. Applications can be divided into two categories: long distance optical communications, and short haul light transmissions such as thermal radiation measurement and laser power transmission.

7.1 Long distance optical communications

7.1.1 Introduction

Since Kapron *et al* (1970) announced a 20 dB km^{-1} silica-based optical fiber, rapid advances have been made in loss reduction of silica-based optical fibers. A transmission loss of as low as 0.2 dB km^{-1} has been obtained after extensive studies on fabrication techniques such as impurity reduction and improvement of glass homogeneity (Miya *et al* 1979). As a result of such progress in silica-based fiber, it became clear that the intrinsic optical loss originates from Rayleigh scattering, the infrared absorption edge and the ultraviolet absorption tail, as shown in section 2.3, and the loss value of 0.2 dB km^{-1} is the ultimate intrinsic loss for a silica-based fiber. It was therefore concluded that in order to reduce the fiber loss further, new infrared fiber materials are required whose infrared absorption edges are located at longer wavelengths than the silica-based glasses.

In 1978, Pinnow *et al*, Van Uitert and Wemple, and Goodman discussed, for the first time, the possibility of ultra-low loss, less than 10^{-2} dB km^{-1}, for infrared transmitting materials, and these discussions motivated research efforts on the non-silica-based infrared fiber materials. Infrared materials for ultra-low fibers are halides, chalcogenides and heavy-metal oxides, as discussed in chapters 4 and 5.

Ultra-low loss optical fibers can be used for long distance optical

196

communications. In particular, nonrepeated intercontinental undersea optical communications are attractive because the existence of the repeaters lowers the reliability of communication.

7.1.2 Properties required for long distance optical communications

Optical fibers used for long distance optical communications must be single-mode optical fibers because of the need for high bit rates in communications. Properties required for such single-mode optical fibers are ultra-low transmission loss and small material dispersion. Of course, it is necessary to consider many other properties, such as splicing and cabling losses. Moreover, stable light sources and detectors operating at longer wavelengths than the conventional optical communications are necessary for constructing optical communication systems.

7.1.2.1 Transmission losses

The achievement of ultra-low loss in an optical fiber depends on several basic material requirements, as shown by Van Uitert and Wemple (1978). These include the following.

(i) The tail of the fundamental electronic absorption edge must lie at sufficiently short wavelengths. This stipulation is easy to satisfy if the optical wavelengths of interest fall well into the infrared.

(ii) The tail of the first lattice absorption edge in the infrared must be at a wavelength well beyond that of interest.

(iii) Scattering losses such as Rayleigh scattering loss must be minimized, implying operation as far in the infrared as possible consistent with other limitations, and choice of a material whose scattering loss is intrinsically small.

(iv) Impurity and defect absorptions must be exceedingly small, i.e. $\leq 10^{-9}$ cm^{-1}. This specification appears to be reasonable for optical wavelengths that lie in a region of the infrared where absorptions of fully oxidized transition metal ions and H_2O are very weak.

(v) The fiber material must be stable both chemically and physically.

Among the above requirements, the tail of the lattice absorption edge and the scattering losses play important roles in determining transmission losses of infrared fibers.

Figure 7.1 shows the calculated loss spectra for fibers made of the following four groups of materials for infrared use: halide, chalcogenide and heavy metal oxide glasses, and a halide crystal. The spectra were calculated by Shibata *et al* (1981) for halide and chalcogenide glasses, by Olshansky and Scherer (1979) for GeO_2 glass, and by Harrington (1981) for KCl crystal. As shown in the figure, the oxide glass has a relatively narrow transmission band and will not exhibit ultra-low loss. Fluoride glass has ultra-low loss potential at the 2–4 μm band and chalcogenide glass, for example germanium–sulfide glass, has a longer cut-off wavelength than the two former groups. Halide crystalline fiber exhibits significantly low loss and a wide transmission band if there are no defects giving rise to additional states within the bandgap, which gives rise to the weak absorption tail.

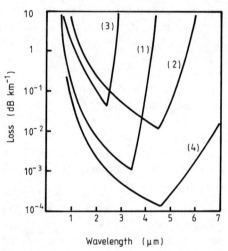

Figure 7.1 Calculated loss spectra for various materials: (1) halide glass (ZrF$_4$–BaF$_2$–GdF$_3$), (2) chalcogenide glass (GeS), (3) heavy-metal oxide glass (GeO$_2$), (4) crystal (KCl) (after Miyashita and Manabe 1982, © 1982 IEEE).

It can be concluded that the most appropriate material for ultra-low loss optical fiber is fluoride glass, although the minimum loss so far reported is around 1 dB km^{-1}, which is one order of magnitude larger than that of the silica optical fiber currently used. However, it should be stressed that research on ultra-low loss optical fibers is still under way, and so further development is possible.

7.1.2.2 Material dispersion

Dispersion is a very important parameter, and is related to the transmission bandwidth of the optical fiber. The material dispersion is defined by

$$M = -\frac{\lambda}{c}\frac{\mathrm{d}^2 n}{\mathrm{d}\lambda^2} \tag{7.1}$$

as shown in section 2.2.2. In equation (7.1), λ is the vacuum light wavelength, n is the refractive index, and c is the velocity of light. When the material dispersion falls to zero, the transmission bandwidth of the single-mode fiber reaches its maximum value. A detailed description of material dispersion in glass and crystalline materials can be found in chapters 4 and 5. A summary of the dispersion properties is as follows.

(i) Of the infrared materials for optical fibers, material dispersion in fluoride glass has been most extensively investigated. It has been determined that the zero material dispersion wavelengths are located at the 1.4–1.8 μm band and the slope of the dispersion versus wavelength is gentle compared with that for silica glass fiber. Furthermore, zero material dispersion wavelengths are somewhat shorter than the predicted minimum loss wavelengths for the fluoride glasses.

It should be noted, however, that waveguide dispersion can compensate material dispersion. Refractive index difference between the core and cladding, as well as the core diameter, relates directly to the waveguide dispersion. Therefore by changing the refractive index difference and the core diameter, the zero total dispersion wavelength can be shifted to longer wavelengths. For example, the optimum core size would appear to be around 12 μm, which would give zero total dispersion at a wavelength of 2.5 μm. At the lower loss wavelength of 3 μm, a total dispersion of about 3 ps Å^{-1} km^{-1} would be estimated. This would allow a system bandwidth of beyond 300 GHz Å^{-1} km^{-1}. Thus, fluoride glasses offer the potential for very large information carrying bandwidths over a wide spectral range (Gannon 1981).

(ii) Material dispersions of chalcogenide glasses fall to zero at longer wavelengths than the fluoride glasses. For example, the zero material dispersion wavelength for As_2S_3 glass is 4.89 μm, which coincides fortunately with the predicted minimum loss wavelength (Miyashita and Manabe 1982). Also the slope of the curve is relatively low compared to oxide and fluoride glasses. Therefore, the chalcogenide glasses are well suited for wide bandwidth optical fibers. However, the achievement of the low loss is, unfortunately, more difficult than in fluoride glasses because of the existence of the weak absorption tail.

(iii) In oxide glasses, the zero material dispersion wavelength falls mainly at 2–3 μm. Fortunately, this wavelength range corresponds to that where the transmission loss is expected to be minimum. Although the achievement of low loss for the oxide glass fibers is difficult, particularly in the fabrication process, the coincidence in the wavelength giving the low loss and zero material dispersion is preferable for long haul and high bit-rate light transmission.

(iv) The zero material dispersion wavelengths for halide crystalline fibers are located at a considerably longer wavelength than those for oxide, fluoride and chalcogenide glass fibers. For example, the zero material dispersion wavelengths for AgCl and TlCl are 5.1 and 6.6 μm, respectively (Nassau 1981).

7.1.2.3 Splicing and cabling losses

In long haul optical fiber communication systems, splicing and cabling losses become an important problem. Jeunhomme (1981) and Okamura and Yamamoto (1983) have discussed the optimum fiber design for long wavelength band operation.

First of all, it should be noted that installing or handling the fiber cable causes a bending loss. Since infrared optical fibers for long distance optical communication are installed in cables, these extra losses inevitably appear. Calculated results (Okamura and Yamamoto 1983) show that the radius of curvature required to maintain low bending loss is increased by increasing the operating wavelength and decreasing the index difference. When the index difference is more than 0.2%, the bending loss is < 0.01 dB rad^{-1} with a radius of curvature of more than 10 cm.

Secondly, microbending, arising inevitably in cabling, is a serious problem, particularly in fabricating ultra-low loss optical fibers. Microbending losses increase with increasing operating wavelength. A high index difference is required to maintain low microbending loss.

Thirdly, splicing losses should be considered because long distance optical communications require a large number of optical fibers which are sequentially connected. Spliced fibers, whose operating wavelength is long, have a wide tolerance for both displacement and separation misalignment, but the effect of angular misalignment is severe.

It can be shown that the loss caused by microbending is very serious. A high index difference is needed to keep the microbending loss small. Uniform bending loss can be neglected when microbending loss is kept sufficiently low. Tokiwa and Mimura (1986) have recently proposed a preferable fiber design after theoretical and experimental investigation on microbending, splicing, coating materials and fiber stranding. Relative refractive index difference of 0.43(0.6)%, core radius of 6.9(8.1) μm, and fiber outer diameter of 150 μm are proposed as the recommended fiber parameters for an operating wavelength of 2.5(3.5) μm.

7.1.3 Summary

At the present stage of the research, fluoride glass optical fibers are the best candidates for transmission lines of long haul and high bit-rate optical communications. Although the minimum loss reported to date is around 1 dB km^{-1}, further loss reduction is expected from new process technologies such as impurity reduction. Moreover, single-mode optical fibers are preferable for long haul and high bit-rate transmission lines because they have the widest bandwidth. Therefore, the material dispersion plays an important role in light transmission. Fortunately, material dispersion for the fluoride glass fiber is relatively low at the wavelength where the transmission loss is expected to be minimum. In addition, careful design of the fiber structure is required to reduce splicing and cabling losses. Microbending loss caused by cabling governs the extrinsic losses. High index difference between the core and cladding is preferable for reducing microbending loss.

7.2 Nuclear radiation-resistant optical fibers

7.2.1 Introduction
One of the most important applications of infrared optical fibers is the light transmission under nuclear radiation. In general, high energy irradiation gives rise to coloration in the ultraviolet to visible region. Since the wavelength range giving the minimum transmission loss for infrared optical fiber is more than 2 μm, the influence of the coloration on the transmission loss is expected to be small. Therefore, this makes it possible to transmit light even under strong high energy radiation.

Such radiation-resistant optical fibers can be used for light transmission in control systems for nuclear power plants, and there may be a variety of applications in the military field.

Research on radiation effects has been restricted to fluoride glass fibers, since they show the most potential for ultra-low loss fibers.

7.2.2 Radiation effects in fluoride glass fibers
Radiation effects on the light transmission properties have been studied mainly for ZrF_4-based glasses (Rosiewicz and Gannon 1981, Ohishi *et al* 1983, 1985, Fisanich *et al* 1985, Tanimura *et al* 1985, Friebele and Tran 1985). The high energy radiation sources used for the experiments are gamma-rays, X-rays and electron beams.

Figure 7.2 shows the incremental transmission loss spectra of bulk fluoride glass samples which were first measured by Rosiewicz and Gannon (1981). In this experiment, ^{60}Co was used as the radiation source, and the measured sample was a $ZrF_4(57.5)$–$BaF_2(33.75)$–$ThF_4(8.75)$ glass block with a thickness of 2 mm. As shown in the figure, the incremental loss on irradiation takes the form of an essentially featureless curve reaching down from the ultraviolet (300 nm) through the visible and falling to zero, within experimental error, by about 2.5 μm. There is another smaller incremental loss hump at about 5 μm. Therefore, this result shows that the region of low incremental loss (2.5–4 μm) coincides with the range of maximum transparency. This is preferable for applications to light transmission under nuclear radiation.

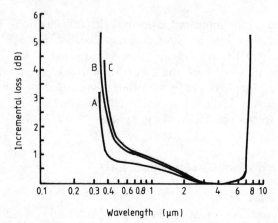

Figure 7.2 Incremental transmission loss spectra of a ZrF_4–BaF_2–ThF_4 glass sample (thickness 2 mm), A, after 1 krad; B, after 7 Mrad; and C, after 45 Mrad (after Rosiewicz and Gannon 1981).

Ohishi *et al* (1983, 1985) have measured the loss increase of fluoride glass fibers directly. The glass fiber compositions measured are $ZrF_4(60.5)$–$BaF_2(31.7)$–$GdF_3(3.9)$–$AlF_3(3.9)$ and $ZrF_4(58.8)$–$BaF_2(30.9)$–$GdF_3(3.7)$–$AlF_3(3.7)$–$PbF_2(2.9)$, which are abbreviated to ZBGA and ZBGAP, respectively. The radiation is ^{60}Co gamma-rays, and the measured fibers are unclad-type Teflon FEP jacket fibers 10 m long. The dose dependences of induced losses for ZBGA and ZBGAP fibers at wavelengths of 2.55 and 3.65 μm are shown in figure 7.3. The induced loss for ZBGAP increases linearly up to 10^6 roentgen, while that for ZBGA tends to saturate at smaller exposures. The induced losses of both ZBGAP and ZBGA become smaller as the wavelength becomes larger. Lead doping largely contributes to radiation softness in the ultraviolet to visible region, but makes the glass radiation hard at wavelengths greater than 3.3 μm. These induced losses are reduced to half of their initial values in 24 hours by thermal bleaching at room temperature.

The doping with some metal ions such as Ti^{4+} and Fe^{3+} changes the formation mechanism of color centers responsible for the induced loss, resulting in a change in the loss characteristics. Ohishi *et al* (1985) reported that Fe- and Ti-doped glasses become considerably more radiation hard, while Ce doping does not have the function of coloration suppression.

It can be concluded from the above results that the induced losses in the 2–4 μm band for ZrF_4-based optical fibers are comparable to those at 1.5 μm for conventional doped silica fibers, which show an induced loss on irradiation by 1 Mrad of 2000 to 30 000 dB km^{-1} at 1.5 μm (Friebele *et al* 1979). However, since studies on the effects of radiation on fluoride glass fibers are still in progress, further improvements are expected in the future. It is possible that certain compositions, when completely purified, may offer resistance to radiation over an extended wavelength region.

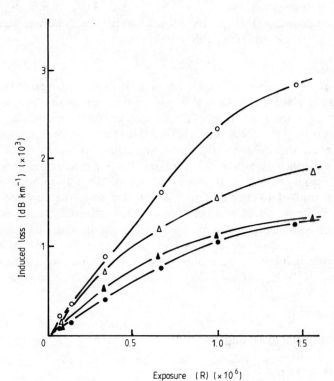

Figure 7.3 The dose dependence of induced transmission losses for ZrF_4–BaF_2–GdF_3–AlF_3 (ZBGA) and ZrF_4–BaF_2–GdF_3–AlF_3–PbF_2 (ZBGAP) fibers at wavelengths of 2.55 and 3.65 μm. The dose rate of gamma rays is 5×10^4 roentgens per hour (after Ohishi *et al* 1983). ZBGAP: $\bigcirc$, 2.55 μm; $\bullet$, 3.65μm. ZBGA: $\triangle$, 2.55 μm; $\blacktriangle$, 3.65 μm.

The mechanism of color center formation by ionizing radiation has been studied extensively by using an electron spin resonance (ESR) method which can identify the color center structure (Fisanich *et al* 1985, Griscon and Tran 1985, Cases *et al* 1985, 1986). The color centers created by ionizing radiation can be classified as two types. One is an electron center identified as Zr^{3+} and Hf^{3+} ions formed by electron trapping at the quadrivalent metal ion site in ZrF_4- and HfF_4-based glasses, respectively. This electron center yields the 463 nm absorption band in the spectrum of ZrF_4-based glass (80 K). The other is a trapped hole center such as F_2^- or F^0, which are responsible for 290 nm absorption of ZrF_4-based glass at 80 K. Some of these hole centers may be associated with oxide impurities. In addition, the addition of Pb to the glass has been found to reduce the Zr^{3+} absorption intensity while introducing bands attributed to Pb^{3+} and Pb^+, confirming the amphoteric ability of Pb^{2+} in the glass to trap both radiolytic electrons and holes (Friebele and Tran 1985).

7.2.3 Summary
At the present stage of research on radiation-resistant infrared optical fibers, the radiation resistance is not necessarily high compared to conventional silica glass fibers. Radiation-induced losses in the ZrF_4-based fluoride glass fibers are 1000–2000 dB km^{-1} in the near-infrared regions when the dose is 10^6 roentgen, which is not as small as in conventional silica glass fibers. Therefore, further research is needed to reduce the induced losses. It should be noted that only fluoride glasses have been studied as fiber materials so far, and since the applications of radiation-resistant fibers are very wide, other materials, such as chalcogenide glasses, should also be studied for use in short haul applications. Finally, hollow waveguides are thought to be ideal radiation-resistant waveguides, because their hollow cores are not damaged by radiation.

7.3 Infrared remote sensing

7.3.1 Introduction
Infrared optical fibers can be applied to various remote sensing systems. Radiometric temperature measurements, infrared image transmissions and remote sensing of infrared spectroscopy are the most promising since it is often necessary to know the temperature, infrared image and infrared spectra in unfavorable environments such as engines, furnaces and nuclear reactors. The use of infrared optical fibers is particularly advantageous because these fibers can transmit longer wavelength light which makes it possible, for example, to measure lower temperatures such as room temperature.

7.3.2 Radiometric temperature measurements
Radiometric temperature measurement systems using optical fibers are basically composed of detecting heads for focusing the radiated infrared light, optical fibers for light transmission, and detectors, as shown in figure 7.4. The infrared light radiated from the object materials has its inherent spectrum that is a function of temperature. The spectrum is expressed by Planck's formula when the radiated body is a black body.

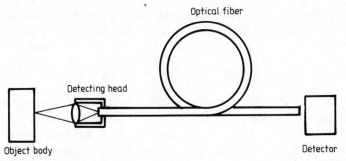

Figure 7.4 A schematic of the apparatus for radiometric temperature measurement using an optical fiber.

Planck's formula is

$$B(\lambda) = C_1\lambda^{-5}/[\exp(C_2/\lambda T) - 1], \tag{7.2}$$

where $B(\lambda)$ is the radiation power per unit wavelength, λ is the wavelength, T is the absolute temperature of the black body, and C_1 and C_2 are constants. Figure 7.5 shows the spectra of the black bodies as functions of wavelength and temperature. Equation (7.2) and figure 7.5 show that the temperature can be derived from the measured spectrum. Strictly speaking, an absolute value of temperature can be obtained from at least two values of radiation power detected at different wavelengths, while only a relative value can be obtained by the detection of the light power at a fixed wavelength. It should be noted that the power detection at longer wavelengths is required for the low temperature measurement, as indicated in figure 7.5. Therefore infrared optical fibers are particularly advantageous for low temperature detection.

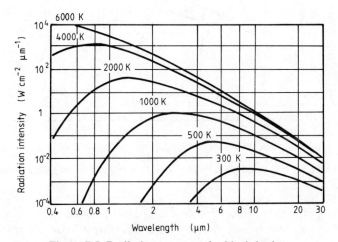

Figure 7.5 Radiation spectra of a black body.

In radiometric temperature measurements using optical fibers, the amount of radiated light power that can be transformed to an electric signal at the detector can be described as (Shimizu and Kimura 1981)

$$E = \pi S(NA)^2 \int f(\lambda, l) \, B(\lambda) \, D(\lambda) \, d\lambda, \tag{7.3}$$

where $B(\lambda)$ is Planck's formula, S is the core area of the fiber, NA is the numerical aperture, $f(\lambda, l)$ is the transmission loss of the fiber, $D(\lambda)$ is the relative sensitivity of the detector, and l is the fiber length.

The detectable minimum temperature was estimated from equation (7.3) by Shimizu and Kimura (1986). In their calculation, the detectable temperature is defined as the temperature giving 10 dB of signal-to-noise ratio. Furthermore, the sensitivity of the detector plays an important role in determining the minimum temperature. Figure 7.6 shows the wavelength dependences of the sensitivities of typical detectors. Shimizu and Kimura (1986) calculated the minimum temperatures for the fiber and detector pairs of silica glass fiber and Ge, fluoride glass fiber and InSb, and TlBr–TlI polycrystalline fiber and HgCdTe. The core diameters and numerical apertures are fixed to be 0.5 mm and 0.2, respectively. Figure 7.7 shows the calculated results, where the detectable regions are inside the lines. As shown, although the minimum detectable temperature for the silica glass fiber is about 200 °C, both the fluoride glass fiber and the TlBr–TlI polycrystalline fiber can measure temperatures below room temperature.

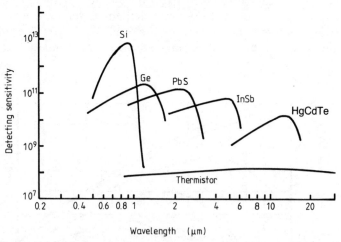

Figure 7.6 Wavelength dependences of sensitivities of typical detectors.

The experiments on radiometric temperature measurement were performed using ZrF$_4$-based fluoride glass fibers and TlBr–TlI polycrystal-

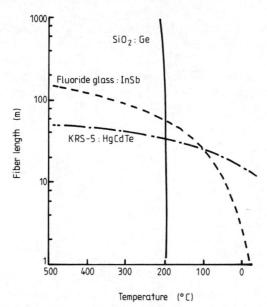

Figure 7.7 Detectable minimum temperatures for various pairs of fibers and detectors (core: 0.5 mm $\varnothing$; *NA*: 0.2) (after Shimizu and Kimura 1986).

line fibers (Shimizu and Kimura 1986). The fluoride glass fiber used is 28 m long and has a 160 μm core diameter, 250 μm outer diameter, and *NA* of 0.14. The bundled fiber with seven fluoride glass fibers shown above was also studied. The length of the bundled fiber is 1.3 m. The TlBr–TlI fiber has a 500 μm core diameter, *NA* of 0.4 and length 4.8 m. A CaF$_2$ lens and InSb detector (77 K) were used for the fluoride glass fiber experiment, while a ZnSe lens and HgCdTe detector (77 K) were used for the TlBr–TlI fiber experiment. A chopping system was adopted for reducing noise. Figure 7.8 shows the results: the curves show the calculated values, which are in good agreement with the experimental values. It can also be seen from the figure that in the case of TlBr–TlI fiber the signal-to-noise ratio is over 15 dB for a temperature of as low as 60 °C. Therefore, the TlBr–TlI polycrystalline fibers are preferable for measuring low temperature. Fluoride glass fibers are also advantageous if the fibers are bundled.

Shimizu and Kimura (1986) fabricated a prototype of the radiometric temperature measurement system. The fiber has a TlBr–TlI core with a 500 μm diameter and TlBr cladding. The fiber length is 4.8 m. This prototype shows a detectable temperature range of 60–300 °C and a temperature resolution of 1 °C.

On the other hand, Katsuyama *et al* (1985) fabricated a radiometric temperature measurement system using a Te-based chalcogenide glass fiber. The Te-based glass fiber can transmit longer wavelength light in the same manner as the TlBr–TlI polycrystalline fiber, and it has become apparent that the measurement of the temperature below 50 °C is possible. In addition, it has been reported by Ishida (1984) that As–S chalcogenide glass fibers can be used for radiometric temperature measurements.

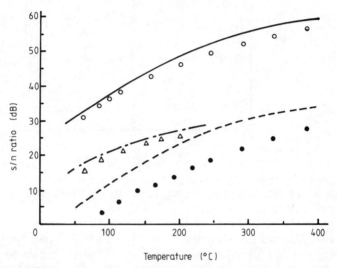

Figure 7.8 Experimental results of signal-to-noise ratio versus temperature (○, ZrF₄ glass fiber (core diameter 160 μm, bundled with seven optical fibers, length 1.3 m); ●, ZrF₄, glass fiber (160 μm, 28 m); △, KRS-5 fiber (500 μm, 4.8 m)) (after Shimizu and Kimura 1986).

7.3.3 Infrared image transmissions

Saito *et al* (1985) fabricated an infrared image guide consisting of 200–1000 optical fibers with As–S glass cores and Teflon FEP claddings. The fabrication procedure is described in section 4.3.3.2. The cross section of the infrared image guide is shown in figure 7.9.

By using this infrared image guide, the thermal image of the candle flame was clearly obtained, as shown in figure 7.10. A feasible use of this device is thus the clear mapping of the temperature distribution of objects far from the observer or located in an inconvenient place.

7.3.4 Infrared spectroscopy

It is often necessary to measure the infrared spectra of various molecules such as CH_4 and C_2H_2 in a chemical analysis. Remote sensing systems are required to detect the gas compositions of some explosive

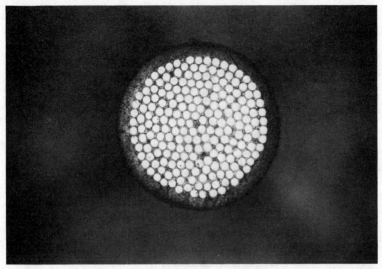

Figure 7.9 A cross-sectional picture of an infrared image guide. This guide is 2.0 mm in diameter, and includes 200 As–S glass fiber cores each 90 μm in diameter (after Saito *et al* 1985).

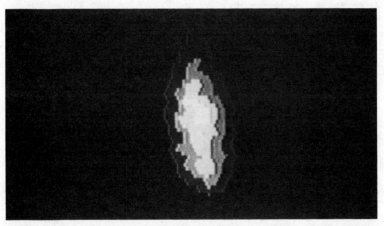

Figure 7.10 A thermal image of a candle flame observed through an infrared image guide (after Saito *et al* 1985).

materials safely. In such cases infrared optical fibers are particularly useful because of their wide transparent wavelength regions. The absorption peaks of various molecules are shown in figure 7.11.

Detecting systems using the infrared optical fibers can be divided into four categories, which are based on the measurement of:

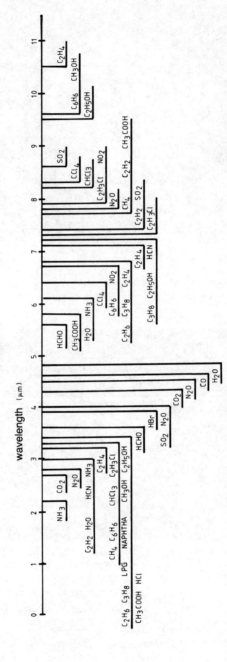

Figure 7.11 Absorption peaks of various molecules (after Saito 1985).

(i) absorption spectra of the transmitted light,
(ii) reflection spectra,
(iii) absorption spectra by using an evanescent wave (ATR method),
(iv) emission spectra.

The apparatus for each method is shown schematically in figure 7.12.

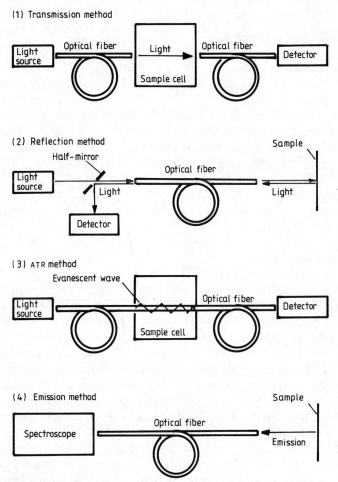

Figure 7.12 Schematics of the apparatus used in various detecting systems (after Saito 1985).

One example of this application was shown by Saito (1985). He measured the absorption intensity of carbohydrates contained in the engine of a car using an As–S chalcogenide glass fiber, and reported that the strong vibration of the engine did not affect the measuring process.

7.3.5 Summary

Radiometric temperature measurement systems using infrared optical fibers are particularly interesting because of their high sensitivities. The detectable temperature range reaches below room temperature. In addition, the temperature profile can be measured if bundled infrared fibers are used.

7.4 Laser power transmissions

7.4.1 Introduction

Since Pinnow *et al* (1978) announced 2 W c.w. CO_2 laser beam transmission through a polycrystalline fiber, power deliveries through optical fibers have become a center of interest in infrared optical fiber research. Crystalline fibers such as TlBr–TlI and Ag halide fibers have been main research targets because their transparent wavelength regions coincide with the CO_2 laser wavelength. Chalcogenide glass fibers and hollow waveguides have also been studied for both CO and CO_2 laser power transmission.

Laser surgery, welding and machining are attractive applications of infrared optical fibers.

7.4.2 Laser surgery
7.4.2.1 General background

Laser applications in medical and surgical fields have been steadily growing. The first successful application was Ar and Nd–YAG laser scalpels using conventional silica glass fibers. Since the operating wavelengths of both Ar and Nd–YAG lasers are below $1.06 \, \mu m$, conventional silica glass fibers can be used for laser power transmission without serious absorption. These scalpels are particularly useful for the coagulation of the blood, because blood efficiently absorbs visible and near-infrared light, resulting in the coagulation by the yielded heat.

On the other hand, when infrared light with a wavelength above $2.5 \, \mu m$ is irradiated to the organism, the light is absorbed by water, which is the primary component of the body. Figure 7.13 shows the relation between the absorption coefficient and the wavelength for water. The operating wavelengths of various lasers are also shown in the figure. It can be confirmed from the figure that the large absorption coefficient arises at wavelengths beyond $2.5 \, \mu m$. Therefore, the irradiation by high power laser light whose wavelength is above $2.5 \, \mu m$ results in the evaporation of the tissue, making possible a new type of scalpel. Both CO and CO_2 lasers (whose wavelengths are 5.3 and $10.6 \, \mu m$, respectively) are promising candidates because of their stable lasing characteristics.

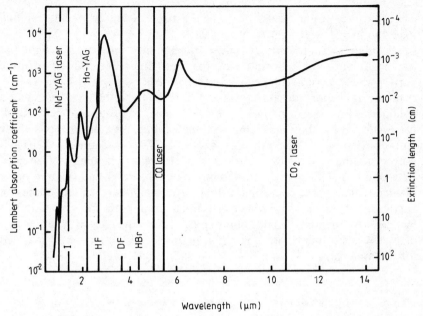

Figure 7.13 The relation between absorption coefficient and wavelength for water (after Arai *et al* 1986).

Chalcogenide glass, polycrystalline and single-crystalline fibers have been experimentally proven to transmit a large amount of CO or CO_2 laser power. Sakuragi *et al* (1981), Ikedo *et al* (1986), Takahashi *et al* (1986), and Mimura and Ota (1982) announced 68 W (TlBr–TlI polycrystalline fiber), 130 W (TlBr–TlI polycrystalline fiber), 50 W (Ag halide polycrystalline fiber), and 47 W (CsBr single-crystalline fiber) transmission of CO_2 laser light, respectively. In addition, Hattori *et al* (1984) reported a 40 W CO laser light transmission through an As–S sulfide glass fiber.

Hollow waveguides such as metallic hollow waveguides can also transmit high power laser light. Although they are less flexible, they are basically advantageous for high power transmission because their hollow cores can dissipate heat easily. For example, more than 200 W c.w. CO_2 laser power can be transmitted through a parallel-plate waveguide (Garmire *et al* 1979).

7.4.2.2 The design concept for the laser scalpel
An optical fiber for surgery use is essentially different from that for optical communication. The transmitting power through the optical fiber for the surgery ranges, for instance, from 20 to 40 W, but light modulation is not essential. Therefore, the energy transmission efficiency is more important than mode characteristics. Moreover, the fiber

cable for the surgery must be designed to dissipate the heat resulting from the high power light transmission.

The fiber cable for surgery must transmit not only a laser light but also a guiding pilot light and a cooling gas. The cable consists of three parts, namely a connector, cable and handpiece.

In the following, design considerations for the cable, connector and handpiece are described. First, the cable must be protected mechanically. Figure 7.14 shows a cross-sectional cable structure proposed by Ishiwatari *et al* (1986). In this cable, a $500\,\mu$m diameter TlBr–TlI polycrystalline fiber with a loose fitting sleeve as cladding is protected by a stainless steel sleeve with 0.6 mm inner diameter and 0.9 mm outer diameter. The cable is designed to give a minimum bending radius of curvature of 15 cm, to protect the fiber from both fracture and loss increase by bending. The outer pipes are covered with an 8 mm diameter jacket made of vinyl chloride and a Teflon sleeve is provided for the inner pipes.

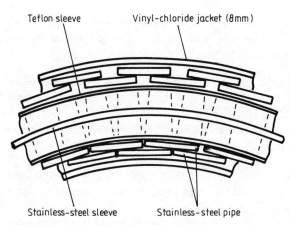

Figure 7.14 The cross-sectional cable structure for limiting bend radius (after Ishiwatari *et al* 1986, © 1986 IEEE).

The cable is composed of an optical fiber for CO_2 laser light, a conventional fiber for pilot light and a channel for cooling gas. Figure 7.15 shows a connector which connects the light and gas to the cable (Ishiwatari *et al* 1986). In this connector, the tip of a TlBr–TlI fiber is placed in a completely sealed space for the protection of the fiber from moisture. The light from a CO_2 laser is focused onto the polished fiber end face through a ZnSe window.

The optical system for the handpiece proposed by Ishiwatari *et al* (1986) is shown in figure 7.16. The high power laser light is focused by a

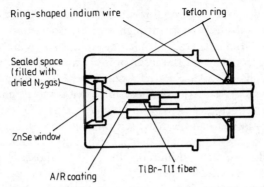

Figure 7.15 The connector for a laser scalpel (after Ishiwatari *et al* 1986, © 1986 IEEE).

ZnSe lens. The pilot light, which is usually He–Ne laser light, is focused onto the appropriate position by the same lens. The lens provided at the tip of the handpiece may become dirty from the smoke and vaporized tissue, so in order to keep it clean, gas is sprayed out of the circumference of the lens and acts as a shield.

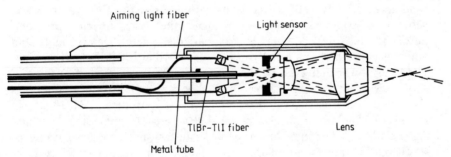

Figure 7.16 The handpiece for a laser scalpel (after Ishiwatari *et al* 1986).

Safety requirements for the application of high power transmission such as in laser scapels incude:

(i) damage to a fiber should be detected immediately and then laser beam irradiation should be stopped automatically,

(ii) the fiber should be totally sealed in order to maintain safety even in the case of the fiber breaking and melting,

(iii) the bend radius must be limited to an acceptable minimum.

7.4.2.3 Experiments with the laser scalpel
First, experimental results of laser power transmission are summarized.

Table 7.1 shows the reported results for the CO_2 laser (wavelength 10.6 μm) power transmission through various fibers and waveguides. It can be seen from the table that the materials mainly studied are TlBr–TlI polycrystals, and the laser power transmitted through the fiber reaches as high as 130 W. Parallel-plate metallic waveguides can also transmit high power laser light whose intensity is more than 200 W.

It has been reported, as shown in table 7.2, that CO laser (wavelength about 5.3 μm) power can be delivered through chalcogenide glass fibers. Chalcogenide glass fibers are particularly advantageous because the transmission losses are considerably lower than those of polycrystalline fibers at the CO laser wavelength. Laser power of 40 W can be transmitted through the fiber, as shown in the table.

Note that the laser power required for a laser scalpel is less than 100 W, which has already been attained, as shown above. This fact has motivated research on laser scalpels using infrared optical fibers, although various problems still remain for such applications. One example is the loss increase due to plastic deformation in polycrystalline fibers. Here, various prototypes of the laser scalpels so far reported are described.

Figure 7.17 shows the laser scalpel constructed by Ishiwatari *et al* (1986). This scalpel consists of a connector, optical fiber cable, and handpiece. A TlBr–TlI polycrystalline fiber used for CO_2 laser power delivery is completely sealed to avoid the fiber toxicity problem. Furthermore, safety assurance for patients, medical personnel, and others is considered, especially in the case of fiber breaking and melting. The laser power delivered by this scalpel ranges from 20 to 40 W.

A laser scalpel using a silver halide polycrystalline optical fiber has also been constructed (reported in Nikkei Electronics 1982 **10** 25, in Japanese). The fiber consists of an AgBr core (1.5 mm diameter) and AgCl cladding (0.35 mm thickness) and can transmit CO_2 laser power of 40–50 W. The advantage of this silver halide fiber is that it is free from the toxicity problems which arise with a TlBr–TlI fiber.

CO laser power transmission has been achieved by using an As_2S_3 chalcogenide glass fiber (Arai *et al* 1986). In this case, a Nd–YAG laser light with a wavelength of 1.06 μm can also be transmitted through the same fiber. Therefore, the coagulation of the blood by the irradiation of a Nd–YAG laser becomes possible, as well as incisions by the high power CO laser light.

Parallel-plate metallic waveguides are also used in the CO_2 laser scalpel (Kubo and Hashishin 1986). The transmitted light power in this case is 81 W and the beam is focused to 0.4 mm in diameter. In order to focus the beam from the parallel-plate waveguide, a cylindrical lens as well as a convex lens must be used because of the flat shape of the light power distribution. The fabricated handpiece is shown in figure 7.18.

Table 7.1 Typical experimental results on CO_2 laser (10.6 μm) power transmission through various optical fibers and waveguides.

Category	Material or structure	Diameter and length	Power (W)	Power density (kW cm^{-2})	Loss (dB m^{-1})	Reference
Polycrystal	TlBr–TlI	0.5 mm	2	–	–	Pinnow et al (1978)
	TlBr–TlI	1 mm, 0.87 m	68	30	0.35	Sakuragi et al (1981)
	TlBr–TlI	0.5 mm, 1.5 m	130	66	0.1	Ikedo et al (1986)
	Silver halide	1 mmϕ	50	6.4	–	Takahashi et al (1986)
Single crystal	CsBr	1 mm, 15 cm	47	6	0.4	Mimura and Ota (1982)
Hollow waveguide	Parallel-plate metallic	–	>200	–	–	Garmire et al (1979)
	Parallel-plate metallic	–	95	40	–	Kubo and Hashishin (1986)
	Circular-metallic	1.2 mm, 80 cm	4	–	–	Nakatsuka and Kubo (1979)

Table 7.2 Typical experimental results on CO laser (5.3 μm) power transmission through various optical fibers.

Category	Material	Diameter and length	Power (W)	Power density (kW cm^{-2})	Loss (dB m^{-1})	Reference
Glass	Ge–As–Se	0.5–0.8 mm, 1–1.5 m	7	4	0.7	Dianov et al (1984)
	As–S	1 mm, 420 cm	40	10	0.3	Hattori et al (1984)
	As$_2$S$_3$	0.2 mm	4	12.8	–	Arai and Kikuchi (1984)
	As$_2$S$_3$	0.4 mm, 1.55 m	15.3	12.2	<1	Arai et al (1986)

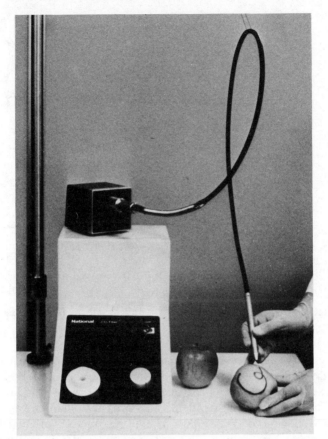

Figure 7.17 A laser scalpel with an optical cable (after Ishiwatari *et al* 1986, © 1986 IEEE).

7.4.3 Summary

Laser surgery is the most attractive application of CO and CO_2 laser power transmission. Laser machining, including laser welding, cutting and incision of some materials such as metals, is also interesting. For such applications crystalline fibers such as TlBr–TlI, silver halide, and CsBr have been studied, together with the hollow waveguides such as parallel-plate metallic waveguides. Chalcogenide glass fibers are also applicable to CO laser power transmission.

7.5 Other applications

7.5.1 Fiber Raman lasers

GeO_2-based oxide glass fibers can be applied to devices using non-linear

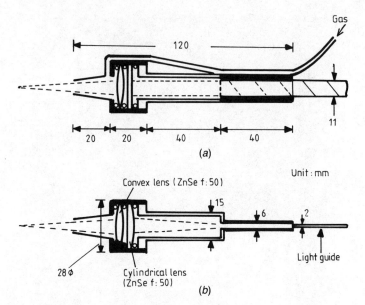

Figure 7.18 The handpiece for a parallel-plate metallic waveguide. (*a*) front view, (*b*) side view (after Kubo and Hashishin 1986).

optical effects, namely fiber Raman lasers that can be used as tunable laser light sources. Since the cross section of Raman scattering of GeO_2 glass is about nine times higher than that of SiO_2 glass, the higher order Raman lines, up to ninth order, can be easily obtained, as shown in figure 4.9 (Sugimoto *et al* 1986). Therefore, tunable laser light sources can be obtained by joining the GeO_2-based glass fibers and high power lasers such as Nd–YAG lasers. In addition, Raman amplification in fluoride glass fibers has also been observed by Durteste *et al* (1985). Raman gain in excess of 40 dB was obtained in an 18 μm diameter core, 10 m long fiber (ZrF_4–BaF_2–LaF_3–AlF_3–NaF) with a pump power in the order of 1 kW.

These light sources are expected to be applied to various techniques for spectroscopic analysis.

7.5.2 Magnetic sensors

TlBr–TlI polycrystalline fibers and As–S chalcogenide glass fibers exhibit the magneto-optical effect (Sato *et al* 1983, 1985). When the magnetic field is applied parallel to the straight fiber, the angle θ of Faraday rotation for linearly polarized light is expressed approximately by

$$\theta = VHL, \tag{7.4}$$

where V is the Verdet constant, H is the magnetic field, and L is the length of the applied field. Equation (7.4) shows that the rotation angle depends linearly on the Verdet constant of the fiber material. Therefore, the magnetic field can be measured by detecting the angle of rotation.

Measured values of the Verdet constants are 5×10^{-4} deg $Oe^{-1} cm^{-1}$ at a $10.6 \mu m$ wavelength for TlBr–TlI polycrystalline fibers and 1.62×10^{-2} min cm $^{-1}$ G $^{-1}$ at 3.39 μm for As–S chalcogenide glass fibers. These magneto-optic effects can be applied to magneto-optical switches, power controllers and so on.

References

Arai T and Kikuchi M 1984 *Appl. Opt.* **23** 3017–19

Arai T, Kikuchi M, Saito M and Sakuragi S 1986 *Tech. Dig. 22nd Symp. in Institute of Electrical Communication, Tohoku University, Sendai, Japan* 124–35 (in Japanese)

Cases R, Alcala R and Tran D C 1986 *J. Non-Cryst. Solids* **87** 93–102

Cases R, Griscon D L and Tran D C 1985 *J. Non-Cryst. Solids* **72** 51–63

Dianov E M, Masychev V J, Plotnichenko V G, Sysoev V K, Balkalov P J, Devjatykh G G, Konov A S, Schipachev J V and Churbanov M F 1984 *Electron. Lett.* **20** 129–30

Durteste Y, Monerie M and Lamouler P 1985 *Electron. Lett.* **21** 723–4

Fisanich P E, Halliburton L E, Feuerhelm L N and Sibley S M 1985 *J. Non-Cryst. Solids* **70** 37–44

Friebele E J, Sigel G H Jr and Gingerich M E 1979 *Fiber Optics* ed B Bendow and S S Mitra (New York: Plenum) pp 355–67

Friebele E J and Tran D C 1985 *J. Non-Cryst. Solids* **72** 221–32

Gannon J R 1981 *Infrared Fibers* (0.8–12 μm) *Proc. SPIE* Los Angeles, CA, pp 62–8

Garmire E, McMahon T and Bass M 1979 *Appl. Phys. Lett.* **34** 35–7

Goodman C H L 1978 *Solid-State Electron. Dev.* **2** 129–37

Griscon D L and Tran D C 1985 *J. Non-Cryst. Solids* **72** 159–63

Harrington J A 1981 *Infrared Fibers* (0.8–12 μm) *Proc. SPIE* Los Angeles, CA, pp 10–15

Hattori T, Sato S, Fujioka T, Takahashi S and Kanamori T 1984 *Electron. Lett.* **20** 811–12

Ikedo M, Watari M, Tateishi F, Fukui T and Ishiwatari H 1986 *Tech. Dig. 22nd Symposium in Institute of Electrical Communication, Tohoku University, Sendai, Japan* 22–9 (in Japanese)

Ishida 1984 *Sensor Technology (Sensa Gijutsu)* **4** 90 (in Japanese)

Ishiwatari H, Ikedo M and Tateishi F 1986 *J. Lightwave Technol.* **LT-4** 1273–9

Jeunhomme L 1981 *Electron. Lett.* **17** 560–1

Kapron F P, Keck D B and Maurer R D 1970 *Appl. Phys. Lett.* **17** 423–5

Katsuyama T, Matsumura H and Kawakami H 1985 *Tech. Dig. Annual Meeting of the Institute of Electronics and Communication Engineers of Japan* 1037 (in Japanese)

Kubo U and Hashishin Y 1986 *Tech. Dig. 22nd Symposium in Institute of*

Electrical Communication, Tohoku University, Sendai, Japan 43–52 (in Japanese)

Mimura Y and Ota C 1982 *Appl. Phys. Lett.* **40** 773–5

Miya T, Terunuma Y, Hosaka T and Miyashita T 1979 *Electron. Lett.* **15** 106–8

Miyashita T and Manabe T 1982 *IEEE J. Quantum Electron.* **QE-18** 1432–50

Mizushima Y, Sugeta T, Urisu T, Nishihara H and Koyama J 1980 *Appl. Opt.* **19** 3259–60

Nakatsuka M and Kubo U 1979 *Tech. Dig. Annual Meeting of Institute of Electrical Engineers of Japan* 392 (in Japanese)

Nassau K 1981 *Bell Syst. Tech. J.* **60** 327–44

Ohishi Y, Mitachi S, Takahashi S and Miyashita T 1983 *Electron. Lett.* **19** 830–1

—— 1985 *IEE Proc.* **132** 114–18

Okamura Y and Yamamoto S 1983 *Appl. Opt.* **22** 3098–101

Olshansky R and Scherer G W 1979 *Tech. Dig. 5th ECOC and 2nd IOOC, Amsterdam, The Netherlands,* 12.5.1–12.5.3

Pinnow D A, Gentile A L, Standlee A G and Timper A 1978 *Appl. Phys. Lett.* **33** 28–9

Rosiewicz A and Gannon J R 1981 *Electron. Lett.* **17** 184–5

Saito M 1985 *Tech. Dig. 1st Workshop on Optical Fiber Sensors* (Japan Society of Applied Physics) 113–20 (in Japanese)

Saito M, Takizawa M, Sakuragi S and Tanei F 1985 *Appl. Opt.* **24** 2304–8

Sakuragi S, Saito M, Kubo Y, Imagawa K, Kotani H, Morikawa T and Shimada J 1981 *Opt. Lett.* **6** 629–31

Sato H, Kawase M and Saito M 1985 *Appl. Opt.* **24** 2300–3

Sato H, Tsuchida E and Sakuragi S 1983 *Opt. Lett.* **8** 180–2

Shibata S, Horiguchi M, Jinguji K, Mitachi S, Kanamori K and Manabe T 1981 *Electron. Lett.* **17** 775–7

Shimizu M and Kimura M 1981 *Tech. Dig. Annual Meeting of the Institute of Electronics and Communication Engineers of Japan,* p 908 (in Japanese)

—— 1986 *Tech. Dig. 22nd Symp. in Institute of Electrical Communication, Tohoku University, Sendai, Japan* pp 142–50 (in Japanese)

Sugimoto I, Shibuya S, Takahashi H, Kachi S, Kimura M and Yoshida S 1986 *Tech. Dig. 22nd Symp. in Institute of Electrical Communication, Tohoku University, Sendai, Japan,* pp 10–21 (in Japanese)

Takahashi K, Yoshida N and Yamauchi K 1986 *Tech. Dig. 22nd Symp. in Institute of Electrical Communication, Tohoku University, Sendai, Japan* pp 30–35 (in Japanese)

Tanimura K, Ali M, Feuerhelm L N, Sibley S M and Sibley W A 1985 *J. Non-Cryst. Solids* **70** 397–407

Tokiwa H and Mimura Y 1986 *J. Lightwave Technol.* **LT-4** 1260–6

Van Uitert L G and Wemple S H 1978 *Appl. Phys. Lett.* **33** 57–9

8 Conclusions

Infrared optical fibers studied to date can be classified, in principle, into two groups: optical fibers with rigid cores and hollow waveguides.

Optical fibers with rigid cores can be then divided into glass fibers and crystalline fibers. Materials for glass fibers are heavy-metal oxides, fluorides and chalcogenides. Typical examples of heavy-metal oxide glass fibers are GeO_2 and GeO_2–Sb_2O_3 glass fibers. The minimum losses of these oxide glass fibers occur around 2–3 μm in wavelength, but in practice the transmission losses are still high because the elimination of impurities is quite difficult. Therefore, heavy-oxide glass fibers have been used as the medium for non-linear effects such as stimulated Raman scattering. For example, GeO_2 glass fibers can be used as wavelength-tunable lasers in combination with high power light sources such as YAG lasers.

Fluoride glass fibers are, on the other hand, the most promising candidates for ultra-low loss optical fibers in long distance optical communication. In particular, ZrF_4-based fluoride glasses exhibit a lesser tendency toward crystallization and higher infrared transparency. The minimum loss can be obtained at 2–4 μm wavelengths. In addition, fluoride glasses are advantageous because they provide a great deal of compositional flexibility which allows them to be tailored to a broad range of properties essential for forming a compatible core and cladding. The zero material dispersion wavelength is less than 2 μm for almost all the fluoride glass fibers. Although this value is slightly different from the wavelength giving the minimum loss, the magnitude of the material dispersion remains small enough in the minimum loss region.

Among the chalcogenide glass fibers, selenide and telluride glass fibers have a wide range transparency, reaching up to around 10 μm in wavelength. However, the predicted minimum losses are higher than those of other glass fibers, such as fluoride glass fibers. Therefore, the main target of research is on CO_2 laser power transmission at 10.6 μm wavelength, which can be used for laser surgery and laser machining.

Radiometric temperature measurement systems using these fibers are also a promising application because of their wide bandwidth transparency. These systems enable us to measure low temperatures, such as room temperature.

$ZnCl_2$ glass fibers have also been studied. However, it was found that $ZnCl_2$ is rather disadvantageous because it shows hygroscopic behavior which makes fiber fabrication difficult.

Crystalline fibers are divided into polycrystalline fibers and single-crystalline fibers. Materials for polycrystalline fibers are typically TlBr–TlI, AgCl and AgCl–AgBr. In general, polycrystalline fibers show low losses at wavelengths above 10 μm. This is useful for the transmission of CO_2 laser power exceeding 100 W. However, it should be noted that the transmission loss gradually increases over a long period of time because of plastic deformation and grain growth, which are accelerated by the existence of water in the atmosphere. Therefore care must be taken to prevent plastic deformation and grain growth by introducing techniques such as heat treatment and doping.

Materials for single-crystalline fibers are almost the same as those for polycrystalline fibers. The advantage of the single-crystalline fibers is that they possess a wide transparent wavelength region ranging from visible to far-infrared. This makes it possible to transmit a visible monitoring light as well as infrared laser power, and so these fibers can be used in laser scalpels with guiding pilot light beams. However, significant loss increase results from plastic deformation caused by repeated bending, and so these fibers must be handled carefully.

Hollow waveguides consist of metallic hollow waveguides and dielectric hollow waveguides. Metallic hollow waveguides have been studied mainly for the CO_2 laser light transmission at 10.6 μm wavelength. The light guiding mechanism of the metallic hollow waveguide is based on grazing reflection at the inner surface of the metallic guiding wall. More than 200 W c.w. CO_2 laser power has been transmitted through parallel-plate metallic waveguides. Furthermore, circular metallic hollow waveguides with dielectric coatings exhibit low loss for the hybrid HE modes. The introduction of dielectric coating layers can also reduce the bending loss. It is noted that the hollow core is effective in dissipating the heat in the power transmission. However, these waveguides are, in general, not flexible, so their application is somewhat limited.

It should be noted, however, that in the abnormal dispersion region the hollow core surrounded by dielectric cladding can transmit light by total internal reflection because the refractive index of the dielectric cladding is lower than unity. GeO_2–ZnO–K_2O glass was found to be suitable for hollow-core fibers which transmit CO_2 laser light, because the abnormal dispersion region coincides with the CO_2 laser wavelength. Although low loss properties have not yet been obtained because of fabrication difficulties, a loss of less than $0.1\,\mathrm{dB\,m^{-1}}$ is expected from the introduction of this glass composition.

Applications of infrared optical fibers fall into two categories: long distance optical communications and short haul light transmission. Fluoride glass fibers are the most promising candidate for long distance applications because of the possibility of ultra-low loss and low dispersion. On the other hand, applications in short haul light transmission require, for example, wide band transparency or low loss for high power laser light transmission. Short haul applications involve nuclear radiation-resistant optical transmission, infrared remote sensing such as temperature measurement by thermal radiation, and laser power transmission such as laser surgery and laser machining. CO_2 laser power transmission is one of the most important applications of infrared optical fibers; chalcogenide glass and crystalline fibers and hollow waveguides can basically be used for this purpose.

Although research into infrared optical fibers is still in progress, it is evident that significant advances in both fiber fabrication and characterization have already occurred. Thus, infrared optical fibers have broken through the initial barriers in the course of their development, and in the near future will be widely applied to the various fields of infrared optics.

Index